Postharvest Technology and Processing of Horticultural Crops

Postharvest Technology and Processing of Horticultural Crops

Editor

Cornelio Losa

scitus academics

Postharvest Technology and Processing of Horticultural Crops

Edited by **Cornelio Losa**

ISBN: 978-1-68117-580-5
Library of Congress Control Number: 2016946435

SCITUS Academics LLC,
616, Corporate Way, Suite 2, 4766,
Valley Cottage, NY 10989

www.scitusacademics.com

Notice

Printed in United States of America on Acid Free Paper

Preface

Fruits and vegetables are very important food commodities. For self-sufficiency and also for processing, export and to meet additional requirements, a lot of emphasis need to be given to reduce post-harvest losses, besides increasing production and productivity of horticultural produces. Processing plays an important role in conservation and effective utilization of these perishable commodities. Many recent innovations in postharvest technology in have been in response to the desire to avoid the use of costly labor and the desire for cosmetically "perfect" produce. These methods may not be sustainable over the long term, due to socioeconomic, cultural and/or environmental concerns. For example, the use of postharvest pesticides may reduce the incidence of surface defects but can be costly both in terms of money and environmental consequences. In addition, the growing demand for organically produced fruits and vegetables offers new opportunities for small-scale producers and marketers. Local conditions for small-scale handlers may include labor surpluses, lack of credit for investments in postharvest technology, unreliable electric power supply, lack of transport options, storage facilities and/or packaging materials, as well as a host of other constraints.

Post-Harvest Technology & Processing of Horticulture Crops will be of immense valuable for research professionals, quality control personnel and postharvest biology students anyone involved in the technology for handling and storing fresh fruits, vegetables, and ornamentals.

Table of Contents

Chapter 1

Tissue biochemical diversity of 20 gooseberry cultivars and the effect of ethylene supplementation on postharvest life

Maria Anastasiadi[a], Paul M. Mwangi[a], José J. Ordaz-Ortiz[a, 1], Sally P. Redfern[b], Mark Berry[b], Monique S.J. Simmonds[c], Leon A. Terry[a]

[a] Plant Science Laboratory, Cranfield University, Bedfordshire, MK43 0AL, UK

[b] Unilever R&D Colworth, Sharnbrook, Bedfordshire, MK44 1LQ, UK

[c] Royal Botanic Gardens, Kew, Surrey, TW9 3AB, UK

HIGHLIGHTS

- Biochemical profile of English grown gooseberries and differences between tissues characterised.
- Effects of exogenous ethylene application prior to storage were investigated.
- 'Scotch Red Rough' and 'Heart of Oak' where some of the cvs with the highest phytochemical potency.
- Total anthocyanin content increased with storage time.

- Exogenous ethylene did not negatively impact the quality of gooseberries

ABSTRACT

The European gooseberry (*Ribes uva-crispa*) is still an understudied crop with limited data available on its biochemical profile and postharvest life. A variety of polyphenols were detected in the skin and flesh of 20 gooseberry cvs, representing mainly flavonol glycosides, anthocyanins and flavan-3-ols. In contrast, gooseberry seeds were for the first time characterised by the presence of considerable amounts of hydroxycinnamic acid glycosides tentatively identified by UPLC-QToF/MS. All cvs examined represented a good source of vitamin C while being low in sugar. Furthermore, the postharvest stability of bioactives was explored by supplementation of exogenous ethylene in air at 5 °C. Results suggest a low sensitivity of gooseberries to ethylene. The overall quality of gooseberries remained stable over two weeks, showing potential for extended bioactive life.

KEYWORDS

Ribes uva-crispa; Ascorbic acid; QToF/MS; Ethylene; Storage; Bioactive life

1. INTRODUCTION

Gooseberries are deciduous shrubs being members of the *Grossulariaceae* family and genus *Ribes* like blackcurrants, redcurrants, whitecurrants and jostaberries (Bordonaba and Terry, 2011). The genus is comprised of more than 150 diverse species with currants and gooseberries being the most popular (Barney and Hummer, 2005). *Ribes uva-crispa*L. (synonym *Ribes grossularia* L.) is a European species and the most prevalent species among the gooseberries found across the world. It is native to United Kingdom, Caucasus Mountains and North West Africa (Barney and Hummer, 2005). The size of gooseberries varies as does their skin colour ranging from green to pink, red, purple, white, and yellow (Hummer and Dale, 2010). The commercial value of gooseberries is limited at present, mainly due to low demand and high cost of production especially during harvesting (Barney and Hummer, 2005 and Dale, 2000). Other influencing factors include prevalence of crop diseases such as powdery mildew (Barney and Hummer, 2005) and lack of high yielding cvs (Pluta et al., 2010).

In recent years there is a rising trend in domestic cultivation of *Ribes* berries both in Europe and other regions (Barney and Hummer, 2005 and Mitchell et al., 2011). Reasons for the increasing interest include small agricultural requirements, resistance to cold winters and the development of improved cvs with better disease resistance, colour, flavour and yield. Quality of gooseberries is primarily based on its visual, textural, organoleptic and nutritional characteristics (Terry et al., 2009). The later attribute has attracted considerable interest over the years, as some bioactive components in berries have been associated with potential health-promoting properties (Folmer et al., 2014 and Wang and Stoner, 2008).

Despite the amount of information on the qualitative and quantitative content of bioactives in berries, the nutritional quality of gooseberries has not yet been sufficiently explored due to limited commercial interest. Only a few gooseberry cvs have been studied thus far in which, polyphenols have been extracted from the fruit as a whole (Jordheim et al., 2007,Määttä-Riihinen et al., 2004, Pantelidis et al., 2007 and Wu et al., 2004). There is limited information about the spatial contribution across tissues in the final polyphenolic content of gooseberries. The main phenolic compounds reported in gooseberries include anthocyanins (Jordheim et al., 2007, Määttä-Riihinen et al., 2004, Pantelidis et al., 2007 and Wu et al., 2004), flavonol glycosides and proanthocyanins (Chiang et al., 2013,Häkkinen et al., 1999a, Mikulic-Petkovsek et al., 2012a and Russell et al., 2009).

The limited commercial value of gooseberries is also depicted in the scarce data available on the postharvest life of gooseberries and the stability of their presumed bioactive components during storage. At present, only a few reports exist on the physical, physiological and biochemical changes occurring in gooseberries during different storage conditions (Harb and Streif, 2004, Kampuse et al., 2015 and Muizniece-Brasava et al., 2015). In addition, a better understanding is needed on the role of ethylene in the postharvest life of gooseberries. Gooseberries are classified as non-climacteric fruits, although they are able to produce ethylene in low amounts (0.035–0.35 ng kg^{-1} s^{-1} at 20 °C) (Cantwell, 2002 and Thompson, 2002). The role of ethylene on *Ribes* berries, however, has not been thoroughly investigated with reports often contradictory regarding their sensitivity (Cantwell, 2002 and McKay and Van Eck, 2006).

The objective of this study was thus two-fold: to explore the biochemical profile of different tissues of a wide selection of gooseberry varieties grown in the UK, and elucidate postharvest changes in biochemistry and quality characteristics for two gooseberry varieties held for 15 days at low temperature with or without application of exogenous ethylene.

2. MATERIAL AND METHODS

2.1. Chemicals

All HPLC and LC–MS grade solvents were obtained from Fisher Scientific (Loughborough UK). (+)-catechin, (−)-epicatechin, procyanidin B1, procyanidin B2, neochlorogenic acid, caffeic acid, sinapic acid, *p*-coumaric acid, quercetin-rutinoside, quercetin-glucoside, isorhamnetin-glucoside, isorhamnetin-rutinoside were purchased from Sigma–Aldrich (Dorset, UK). Cyanidin-3-glucoside and cyanidin-3-rutinoside, were purchased from Extrasynthese (Genay Cedex, France). Metaphosphoric acid (Bioxtra ≥ 33.5%), L-ascorbic acid and D-fructose were obtained from Sigma–Aldrich (Dorset, UK). D-glucose and sucrose, were purchased from Fisher Scientific (Loughborough UK).

2.2. Plant material and sample preparation

Gooseberry fruits from 20 cvs of *R. uva-crispa* were obtained at optimum maturity, from The National Fruit Collection (Brogdale, Kent, UK) on the 13th of July 2012 for biochemical analysis (Fig. 1). Based on the biochemical results obtained, two cvs 'Careless (Kent)' and 'Scotch Red Rough' were selected and harvested again the following year (12th of July 2013) for the purposes of the postharvest trial. Approximately 100–200 berries were harvested from two plants per cv at optimum maturity stage and transported to Cranfield University in cool boxes within 2 h from collection and immediately snap-frozen in liquid nitrogen. The samples were divided into triplicates and stored at −80 °C before analysis. Each sample was further divided into two subsamples. Half of the material (approximately 100 g) was freeze-dried, the seeds were manually removed and the berries were ground into a fine powder for the extraction of phenolics and sugars. The second subsample was kept fresh frozen and powdered in a mortar grinder (RM 200, Retsch Ltd., Derbyshire, UK) with liquid N_2 and used for the extraction of ascorbic acid to avoid potential degradation of ascorbic acid during the freeze-drying process.

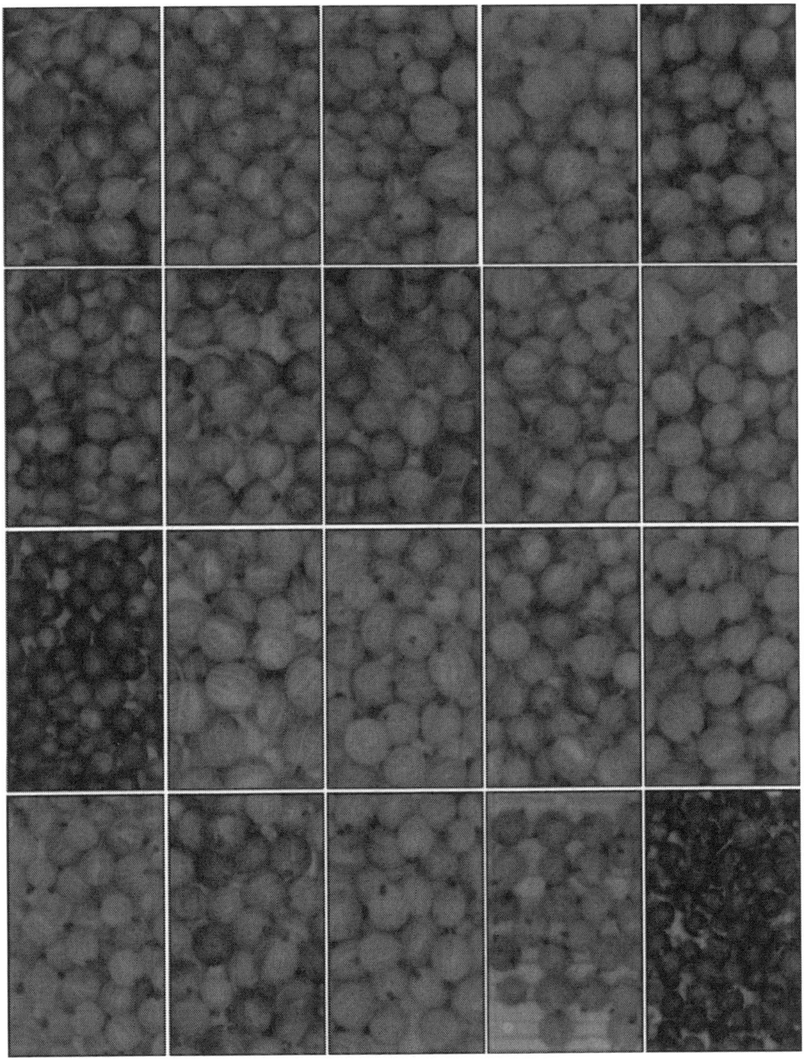

Figure 1. Phenotypic differences of 20 gooseberry cvs, harvested from the National Fruit Collection in 2012. From left to right: 'Careless 3024', 'Guy seedling', 'Jolly Amylers', 'Jubilee Careless', 'Heart of Oak', 'Victoria', 'Ajax', 'Goutray's Earliest', 'Rubuste Nool', 'Lord Elco (Scotland)', 'May Dulle', 'White Eagle', 'Careless (Kent)', 'Nailer', 'Mitre', 'Cousen's Seedling', 'Lord Audley', 'Careless VT 512', 'Bedford Yellow', 'Scotch Red Rough'.

2.3. Postharvest trial

Samples from the two selected gooseberry cvs ('Scotch Red Rough' and 'Careless (Kent)') were transferred inside polystyrene boxes with ice-packs, from The National Fruit Collection to Cranfield within 3 h from harvest. Upon arrival at Cranfield, gooseberries from each cv were split into two batches. The four batches of gooseberries were placed in plastic stackable crates inside water sealed, air-tight polypropylene chambers (88 cm × 59 cm × 59 cm) fitted with two 8 × 8 cm electric fans (Nidec beta SL, RS Components Ltd., Northants, UK) to circulate the ethylene gas during treatment. Two boxes per cv were injected with 11.69 μg L^{-1} of standard ethylene gas (100% ethylene; SIP analytical) and the other two boxes were used as control samples (untreated). Temperature of the storage room was set at 5 °C and the treatment time was 24 h. The concentration of the ethylene gas in the chambers was confirmed after 30 min of injection and after 24 h of storage by withdrawing air from the chambers (including controls), using a tapped 20 mL plastic syringe. The headspace of the sampled air was injected into a gas chromatograph (GC—Model 8340, DP800 integrator, Carlos Erba Instruments, Herts, UK, analytical column, Porapak) fitted with flame ionisation detector (FID) and ethylene gas present quantified.

After the treatment was completed (24 h), every batch was split into three replicates and stored in ventilated propylene (15 cm × 22 cm × 8 cm) containers. The containers had an inlet and outlet that provided ventilation to the fruits by way of pumping air into the boxes at a flow rate of about 3.33 mL s^{-1} from a flowmeter controlled unit attached to a ICA 6000 (International Controlled Atmosphere Ltd., Kent, UK). The RH inside the boxes was maintained at 80 ± 10% by placing a beaker of water inside the containers. The RH and temperature in the boxes was continuously monitored using Tinytag Ultra 2 TGU-4500 data loggers (−95% RH, −25 to 85 °C, Gemini Data Loggers Ltd., West Sussex, UK).

2.3.1. Sampling during storage period

Upon arrival (Day 0), five gooseberries were randomly selected from each cv in triplicate. After the treatment (Day 1), five gooseberries were sampled randomly from each box (3 boxes per treatment). Sampling was thereafter repeated at regular intervals (Day 4, Day 7, Day 11, Day 13, Day 15). All samples were subjected to colour and ethylene production measurements before being snap-frozen in liquid nitrogen. The snap-frozen berries were stored at −80 °C until analysis. Prior to extraction the plant material was freeze-dried and powdered as before and analysed for individual soluble sugars and phenolics. Water content of gooseberries was calculated by recording the weight of all samples before and after freeze-drying.

2.4. HPLC-ELSD analysis of non-structural carbohydrates

Extraction of non-structural carbohydrates was performed according to a previous method with slight modifications (Terry et al., 2007). Prior to analysis, the sugars extracts were diluted (1:9, v/v) with HPLC grade water and injected into an Agilent 1200 HPLC fitted with a prevail carbohydrate ES 5 µm size of 250 nm × 4.6 mm diameter and a guard column of the same type. The mobile phase consisted of solvent A (water) and solvent B (acetonitrile) and the elution gradient was as follows: 0–15 min, 80–50% B, 15–20 min, 50–20% B, 20–25 min, 80% B. The eluted compounds were detected by evaporative light scattering detector (ELSD) and quantification was based on external calibration curves of commercial standards.

2.5. Phenolic compounds

2.5.1. Extraction of phenolic compounds

Extraction of phenolic compounds from berries and seeds was performed according to a previous method with slight modifications (Giné Bordonaba and Terry, 2008). Freeze–dried berry powder (150 mg) and freeze–dried seed powder (50 mg) were extracted with 3 mL and 1 mL, respectively, of methanol/water/HCl (70:29.5:0.5 v/v) in a shaking water bath at 20 °C for 1.5 h. The samples were vortexed every 10 min to re-suspend the solids. The extracts were subsequently filtered with 0.20 µm nylon filters and stored at −40 °C before analysis.

2.5.2. HPLC-DAD analysis of phenolic compounds and fraction collection

Analysis of flavan-3-ols, dimeric procyanidins, flavonol glycosides, anthocyanins and neochlorogenic acid in gooseberry extracts (flesh/skin) was performed on a Agilent 1200 HPLC system with a diode array detector (DAD) detector, equipped with a quaternary pump, a fraction collector, a thermostated column compartment operating at 30 °C and a cooled autosampler set at 6 °C. Separation was achieved with an Eclispse XDB-C18 column (150 mm × 4.6 mm; 5 µm, Agilent Technologies) and a 1 mm OPTI-Guard column (mobile phases consisted of 3% formic acid in water) (solvent A) and acetonitrile (solvent B) and the flow rate was 17 µL s^{-1}. The elution gradient was as follows: 0–5 min, 3–5% B, 5–10 min; 5–10% B, 10–20 min, 10–30% B; 20–25 min, 30–60% B; 25–26 min, 60–100% B, 26–30 min, 100% B, followed by 5 min re-equilibration time. The DAD detector was operating at 4 different wavelengths, 280 nm, 320 nm, 360 nm and 520 nm. Quantification was based on external calibration curves of commercial standards. Six anthocyanins detected at 520 nm for which commercial standards were not available, were quantified based on the calibration curve obtained for cyanidin-3-glucoside. Furthermore the fraction collector was used to isolate those peaks and subject them to LC/MS analysis for further elucidation.

2.5.3. UPLC/QToF/MS analysis of unknown anthocyanins

The six fractions isolated at 520 nm with UV spectra similar to the anthocyanin standards, were further analysed with an Agilent 1290 infinity UPLC system, comprised of a binary pump with a jet weaver V35 mixer, a thermostated column, set at 30 °C, and a cooled autosampler set at 6 °C. The UPLC system was coupled with an Agilent 6540 Ultra High Definition (UHD) Accurate Mass Q-TOF–MS system (Agilent Technologies) equipped with an electrospray ionization source (Agilent Dual Jet Stream). The fractions were evaporated using a miVac centrifugal vacuum concentrator (Genevac) and reconstituted with 100 µL 0.1% formic acid in water and 5 µL were injected into a ZORBAX SB-C18 Rapid resolution, HD 2.1 × 150 mm, 1.8 µm column (Agilent Technologies). Mobile phases consisted of 0.1% formic acid in water (solvent A) and 0.05% formic acid in acetonitrite (solvent B) and the flow was 6.67 µL s^{-1}. The elution gradient was as follows: 0.5–6.5 min, 5% B, 6.5–8 min; 5–15% B, 8–15.5 min, 15–20% B; 15.5–15.55 min, 20–45% B; 15.55–17.55 min, 45–100% B, 17.55–17.60 min,

100–5% B; 17.60–20.0 min, 5% B. Source conditions for the MS were set as follows: nebulizer gas (N_2) temperature 150 °C, at a flow rate of 0.13 L s^{-1} sheath gas (N_2) temperature 400 °C, at a of flow rate of 0.18 L s^{-1} and fragmentor voltage 165 V. MS spectra acquired in full scan, positive ionisation mode in the *m/z* 50–1500 range.

MS/MS experiments were subsequently carried in positive mode; to obtain the daughter ion spectrum of the selected precursor ions. Product ions were produced by collision-induced dissociation using targeted MS/MS experiments, at three different collision energies (10, 20, 40 eV) with an isolation window of 4 *m/z* (medium) for all compounds. The LC-MS system and data acquisition were controlled by an Agilent MassHunter Data Acquisition software B.04.00.

2.5.4. UPLC/QToF/MS analysis of hydroxycinnamic acid derivatives

The gooseberry extracts obtained from different tissues (skin/fresh and seeds) were further analysed with the Agilent 1290 infinity UPLC system and the Agilent 6540 UHD Accurate Mass Q-TOF-MS system described above. All extracts were further diluted (1:4) with 0.1% formic acid in water and 5 μL were injected into a ZORBAX SB-C18 column (Rapid resolution, HD 2.1 × 150 mm, 1.8 μm; Agilent Technologies) with the chromatographic method described in Section 2.5.3. Commercial standards of different phenolic compound groups were used as reference compounds (hydroxycinnamic acids, flavonols, flavan-3-ols, dimeric procyanidins).

MS/MS experiments in positive mode were subsequently carried on selected precursor ions to assist with compound identification. Product ions were produced by collision-induced dissociation using targeted MS/MS experiments, at three different collision energies (10, 20, 40 eV) with an isolation window of 4 *m/z* (medium) for all compounds.

Quantification of the hydroxycinnamic acid derivatives identified with the above method was based on external calibration curves of the corresponding aglycons (*p*-coumaric acid, caffeic acid, ferulic acid, sinapic acid).

2.6. Extraction and quantification of total ascorbic acid

The total ascorbic acid of the 20 gooseberry cvs harvested in 2012, was determined as total vitamin C using a previously described method with slight modifications (Planchon et al., 2004). The method determined the total vitamin C in the gooseberries as the sum of ascorbic acid and its oxidised form dehydroascorbic acid. Briefly, 500 mg of fresh frozen gooseberries were mixed with 2.5 mL of degassed metaphosphoric acid solution (0.01 M) and the mixture was extracted in a shaking water bath at 25 °C for 10 min. The extract was filtered through 0.20 µm nylon filters and immediately injected in the HPLC. Ascorbic acid was monitored at 248 nm and the total vitamin C was quantified using an external calibration curve of L-ascorbic acid. The assay was performed in triplicate.

2.7. Colour measurement

The colour of gooseberries was assessed using an 8 mm aperture handheld tristimulus colourimeter (Minolta CR- 400, Osaka, Japan). The mean colour space coordinates of Lightness (L^*), Chroma (C^*) and Hue angle ($H°$) were taken from three positions on each gooseberry for every sampling point.

2.8. Ethylene production measurement

The ability of the gooseberries to produce ethylene gas was measured with an ETD-300 Ethylene Detector system (Sensor Sense BV, Nijmegen, The Netherlands). In order to determine the amount of ethylene in the cuvettes after 1 h incubation, air was pumped at a continuous flow rate of 1.1 µL s^{-1}. The air from the cuvettes was passed through $CO_2$and water scrubbers. Measurements from every cuvette were made for a period of 12 min and expressed in ng kg^{-1} s^{-1}. The equipment was equilibrated before measurement by measuring an empty cuvette for 1 h. A baseline measurement of 3 min was maintained between the samples to eliminate cross contamination.

2.9. Statistical analysis

Analysis of variance (one-way and two-way ANOVA) was performed using Genstat for Windows, Version 14 (VSN International Ltd., Herts., UK). The differences between means of data were compared through Least Significance Difference (LSD) and they were considered to be statistically significant at the 95% confidence level *($p \leq 0.05$)*. Logarithmic transformations were employed where needed in order to ensure the assumption of equal variability.

3. RESULTS AND DISCUSSION

3.1. Phenolic composition of gooseberries

3.1.1. Characterization and quantification of non-coloured compounds in skin/flesh

Gooseberries (flesh/skin) contained a variety of phenolic compounds including flavonol glycosides, flavan-3-ols, dimeric procyanidins, and hydroxycinnamic acid derivatives (Table 1 and Table 2). The major class of polyphenols present in all cvs was flavonol glycosides with the rutinosides of quercetin and isorhamnetin being the predominant glycosides. Quercetin and isorhamnetin glucoside on the other hand were found in much lower amounts. Mikulic-Petkovsek et al. (2012a) also observed that quercetins in gooseberries existed as glucosides and rutinosides and that these are the main form in which they exist in berries from *Glossulariaceae* family. The same authors have also reported high concentration of isorhamnetin-3-rutinoside in gooseberries (73.9 mg kg^{-1}in red and 48.3 mg kg^{-1} in white gooseberries for FW) when compared to other common berries such as bilberry, blackcurrant and strawberry. Similarly, Määttä-Riihinen et al. (2004) have also reported high concentration of isorhamnetin glycosides in gooseberries. Our results are in good agreement with these previous reports, although the concentration range of quercetin glycosides among the 20 gooseberry cvs analysed within this study were higher than those previously reported (Häkkinen et al., 1999a,Häkkinen et al., 1999b, Määttä-Riihinen et al., 2004 and Mikulic-Petkovsek et al., 2012a). These differences could reflect that, to date, only a limited number of gooseberry cvs have been studied. However, the overall trend is in line with Määttä-Riihinen et al. (2004)observation that quercetins are high in berries from *Glossulariaceae* family. Other flavonol glycosides reported in gooseberries, include glycosides of myricetin and kaempferol, typically present in the form of rutinoside representing minor constituents of the flavonol content of gooseberries (Määttä-Riihinen et al., 2004 and Mikulic-Petkovsek et al., 2012a). Furthermore Mikulic-Petkovsek et al. (2012a), identified the presence of small amounts of lacitrin and syringetin rutinoside in a white gooseberry cv and other berry species.

Flavan-3-ols and dimeric procyanidins were present in gooseberries in small to moderate amounts with (+)-catechin being the predominant with an average concentration of 28.0 mg kg^{-1} (FW). These results are comparable with previous reports, with the exception of (+)-catechin which was in concentrations higher than that previously reported (Häkkinen et al., 1999b). Very few reports are available describing the levels of procyanidins in gooseberries, however, according to Wu et al. (2004), procyanidins and prodelphinidins are present in gooseberries mainly in polymeric form.

Of the phenolic acids, only hydroxycinnamic acid derivatives were detected in low to moderate amounts in the skin and flesh of gooseberries. The presence of neochlorogenic acid (*trans*-5-*O*-caffeoylquinic acid) was verified through the use of a commercial standard and accurate mass measurements. Several hydroxycinnamic acid glycosides were also tentatively identified with QToF/MS/MS, based on accurate mass measurements and the fragmentation pattern acquired with MS/MS in three different voltages (10, 20, 40 eV) which matched the fragmentation pattern obtained for the corresponding aglycons. The most abundant fragments for each compound are presented in Table 3. The sugar moiety was a hexose and the most abundant pseudo-molecular ion in positive mode was in the form of a $[M - Na]^+$ adduct. The following glycosides were identified in gooseberry skin/flesh extracts: *p*-coumaric, caffeic, ferulic and sinapic acid hexosides (Table 3).

Table 1. Flavonoid composition (mg kg^{-1}) of the flesh/skin of 20 gooseberry cvs on a fresh weight (FW) and dry weight (DW) basis (values inside brackets).

Flavonoid composition (mg kg^{-1}) of the flesh/skin of 20 gooseberry cvs on a fresh weight (FW) and dry weight (DW) basis (values inside brackets).[a]

Cultivar	(+)-Catechin		(−)-Epicatechin		ProcB1[b]		ProcB[c]		Q-3-rut[d]		Q-3-gluc[e]		Isorh-3-rut[f]		Isorh-3-gluc[g]	
Scotch Red Rough	20.1	(112.8)	4.2	(23.7)	8.8	(49.5)	2.5	(14.3)	102.5	(767.9)	13.7	(102.7)	81.0	(606.4)	2.7	(20.4)
Lord Audley	26.5	(224.8)	2.2	(18.7)	9.3	(79.2)	2.4	(20.2)	78.2	(585.6)	7.0	(52.8)	38.3	(287.2)	1.6	(12.3)
Heart of Oak	44.3	(264.5)	2.4	(13.9)	12.9	(75.4)	2.3	(13.6)	55.3	(414.1)	7.1	(52.8)	26.8	(200.4)	1.2	(8.7)
Nailer	37.0	(219.3)	2.2	(13.0)	13.1	(70.3)	2.4	(14.4)	49.9	(373.8)	4.8	(35.9)	29.3	(219.1)	1.4	(10.1)
Jolly Amylers	56.5	(364.0)	2.4	(15.6)	16.4	(105.4)	2.3	(15.0)	25.8	(192.9)	3.4	(25.3)	24.0	(179.6)	1.7	(12.7)
Bedford Yellow	46.6	(326.2)	2.7	(18.5)	18.4	(128.5)	4.5	(31.2)	37.5	(280.6)	2.9	(21.9)	13.9	(104.0)	0.9	(6.6)
Cousen's Seedling	14.0	(112.6)	2.0	(16.1)	10.3	(82.5)	3.9	(31.4)	44.7	(334.5)	5.0	(37.3)	40.6	(304.2)	2.5	(18.6)
Careless (Kent)	10.9	(83.2)	1.0	(7.4)	7.0	(53.7)	1.0	(7.8)	43.1	(322.8)	5.3	(39.9)	28.7	(215.0)	2.3	(17.0)
Victoria	34.6	(285.2)	2.3	(18.8)	12.5	(102.2)	2.6	(21.0)	25.7	(192.4)	3.0	(22.5)	16.4	(123.2)	0.5	(3.9)
White Eagle	32.6	(209.6)	2.1	(13.4)	12.4	(79.3)	2.6	(16.8)	16.2	(121.4)	3.3	(25.1)	24.5	(183.2)	2.2	(16.6)
Jubilee Careless	28.5	(192.9)	1.7	(11.6)	10.5	(71.2)	1.6	(10.9)	24.6	(184.1)	2.5	(18.6)	19.5	(145.9)	1.2	(9.3)
Goutray's Earliest	29.6	(258.0)	3.0	(25.8)	10.4	(90.9)	3.7	(32.1)	20.6	(154.4)	3.0	(22.4)	18.1	(135.2)	1.4	(10.6)
May Dulle	15.1	(106.9)	2.6	(18.5)	9.7	(68.6)	1.9	(13.7)	30.6	(229.5)	5.8	(43.5)	21.4	(160.0)	1.5	(11.0)
Mitre	17.8	(126.5)	1.3	(09.5)	8.2	(58.2)	1.2	(8.9)	33.9	(254.0)	3.7	(27.6)	18.8	(141.1)	1.2	(8.9)
Lord Elco	40.1	(320.3)	1.4	(10.9)	13.0	(103.3)	1.7	(13.6)	9.6	(71.6)	0.6	(4.1)	18.3	(137.3)	1.4	(10.9)
Rubuste Nool	30.1	(206.1)	1.6	(10.9)	11.1	(76.0)	1.5	(10.0)	18.1	(135.9)	1.9	(14.5)	18.3	(137.3)	1.4	(10.3)
Careless 3024	26.9	(201.5)	2.3	(17.0)	10.7	(80.4)	2.4	(17.9)	20.2	(150.9)	1.4	(10.6)	15.0	(112.1)	0.6	(4.4)
Careless (VT)	19.9	(130.7)	1.5	(9.7)	8.3	(55.2)	1.5	(10.1)	25.4	(190.3)	2.6	(19.4)	16.8	(125.8)	1.2	(8.8)
Ajax	19.4	(132.1)	1.7	(11.5)	10.3	(69.8)	1.9	(13.2)	19.4	(145.6)	2.3	(16.9)	16.1	(120.8)	1.3	(9.6)
Guy seedling	8.8	(44.7)	1.8	(9.1)	8.7	(43.8)	3.4	(17.2)	15.4	(115.1)	2.6	(19.2)	29.0	(217.4)	2.0	(15.1)
Mean	28.0	(196.1)	2.1	(14.7)	11.1	(77.2)	2.4	(16.7)	34.8	(260.9)	4.1	(31.2)	26.0	(192.8)	1.5	(11.3)
LSD[h]	8.25	(53.48)	0.49	(2.99)	2.27	(12.33)	0.48	(2.77)	16.34	(122.38)	2.10	(15.9)	8.40	(68.4)	0.50	(3.2)

[a] The results are presented in descending order according to the sum of flavonoids for FW.
[b] ProcB1 = procyanidin B1.
[c] ProcB2 = procyanidin B2.
[d] Q-3-rut = quercetin-3-rutinoside.
[e] Q-3-gluc = quercetin-3-glucoside.
[f] Isorh-3-rut = isorhamnetin-3-rutinoside.
[g] Isorh-3-gluc = isorhamnetin-3-glucoside.
[h] LSD = least significant difference at the 95% confidence level.

Among the hydroxycinnamic acid glycosides detected in gooseberry flesh/skin extracts, coumaric and caffeic acid glycosides were the most abundant, followed by ferulic acid glycosides (Table 2). Sinapic acid glycosides were also detected in trace amounts (Table 3). 'Guy Seedling' was the cv with the highest content in hydroxycinnamic acid glycosides with a sum of 114.8 mg kg^{-1} (FW), while 'Scotch Red Rough' cv which exhibited the highest flavonoid content among all cvs, had low levels of hydroxycinnamic acid glycosides (56.1 mg kg^{-1}). The levels of hydroxycinnamic acid glycosides reported here are in good agreement with those reported in berries from other species in the*Grossulariaceae* family (Määttä-Riihinen et al., 2004). Neochlorogenic acid was present in small amounts in gooseberry flesh/skin with a mean concentration of 5.6 mg kg^{-1} (FW) in 'Scotch Red Rough'. Its concentration was almost 2-fold higher in coloured varieties when compared to green varieties. This is the first report of neochlorogenic acid in gooseberries. The presence of several phenolic acids has been previously reported in the literature including protocatechuic acid, *p*-hydroxybenzoic acid, *p*-coumaric acid, ellagic acid, ferulic acid, chlorogenic acid, sinapic acid, caffeic acid, syringic acid, vanillic acid and gallic acid (Filipiak-Szok et al., 2012, Mattila et al., 2006 and Russell et al., 2009). It is notable that none of the above phenolic acids were detected in free form in gooseberries in our study and no hydroxybenzoic acids were detected in the berries of the gooseberry cvs analysed. Similarly Määttä-Riihinen et al. (2004) reported the presence of hydroxycinnamic acids in gooseberries and other berry species in conjugated form, mainly as caffeoyl- and *p*-coumaroylglucosides.

3.1.2. Phenolic profile of gooseberry seeds

Unlike the flesh/skin extracts which contained a variety of different phenolic compounds, the only type of phenolic compounds detected in gooseberry extracts of seeds were hydroxycinnamic acid derivatives, which were tentatively identified as glycosides of caffeic, coumaric, ferulic and sinapic acids as described in Section 3.1.1. Two distinct peaks with the same spectral characteristic attributed to coumaric acid glycosides and ferulic acid glycosides were present in all seed extracts, indicating the presence of two isomers of coumaric acid and ferulic acid glycosides (Table 3). Their concentrations are presented in Table 4. Gooseberry seeds were observed as being particularly high in phenolic acid glycosides, with concentrations being 10 times higher compared to the flesh/skin extracts. There was some variability among different varieties with the highest concentrations observed for 'Heart of Oak'. The qualitative profile was similar to the flesh/skin extracts, with coumaric and caffeic acid glycosides being the most abundant, followed by ferulic and sinapic acid glycosides. No phenolic acids were present in free form in the seed extracts, as it was confirmed by comparison with the corresponding aglycon standards. It is also notable that neochlorogenic acid, which was detected in the flesh/skin extracts, was absent in gooseberry seeds. Other compounds detected in the gooseberry seed extracts included trace amounts of (+)-catechin and (−)-epicatechin, quercetin-3-rutinoside, isorhamnetin-3-rutinoside and cyanidin-3-rutinoside and cyaniding-3-glucoside in some red cvs.

To our knowledge no other reports exist on the phenolic composition of gooseberry seeds, but some comparisons can be made with other *Ribes* species. Gođevac et al. (2012) have tentatively identified 35 compounds in black, red and white currant seed extracts, including phenolic acids derivatives such as sinapoyl glucoside and chlorogenic acid and several phenolic acids in free form, small amounts of flavonols and flavonol glycosides as well as two nitrile containing derivatives of *p*-coumaric and ferulic acid. In contrast Bakowska-Barczak et al. (2009), have reported flavonols as the main phenolic group present in blackcurrant seed residue in the form of kaempferol, myricetin and quercetin glycosides. The same authors also reported the presence of amounts of *p*-coumaric acid along with small amounts of *p*-coumaroyl-glucoside in the seeds (0.487 g kg^{-1} and 0.179 g kg^{-1}, respectively, for DW), although the presence of the aglycon was possibly a degradation product (Bakowska-Barczak et al., 2009). Lu et al. (2002) isolated six glucosides of quercetin, kaempferol and myricetin from blackcurrant seeds along with two nitrile containing derivatives of *p*-coumaric acid. Furthermore, Lu and Foo (2003) isolated and characterised by ^{13}C NMR two phenolic acid conjugates namely 1-*p*-coumaroyl-β-D-glucopyranoside and 1-cinnamoyl-β-D-glucopyranoside from blackcurrant seeds along with several anthocyanins and flavonoids.

The presence of phenolic acid glycosides has also been reported in the genus *Fragaria*, with several hexoses of coumaric, ferulic, caffeic and sinapic acids reported primarily in the strawberry receptacle and in small amounts in the achenes (Fait et al., 2008). These differences in the diversity of compounds and their concentrations could be attributed to differences among species, cvs, extraction and detection methods.

Table 2. Hydroxycinnamic acid derivative composition (mg kg^{-1}) of the flesh/skin of 20 gooseberry cvs on a FW and DW basis (values inside brackets).[a]

Cultivar	Coumaroyl-hexose1		Coumaroyl-hexose2		Caffeoyl-hexose		Feruloyl-hexose1		Feruloyl-hexose2		Neochlorogenic acid	
Guy Seedling	45.5	(229.8)	15.3	(77.3)	39.9	(201.6)	2.3	(11.9)	11.8	(5.97)	10.4	(52.6)
Heart of Oak	41.5	(241.5)	18.2	(105.8)	33.4	(194.2)	1.5	(8.8)	8.2	(4.93)	7.7	(45.4)
Nailer	33.3	(198.8)	12.8	(76.5)	33.5	(199.8)	1.6	(9.9)	4.7	(2.88)	9.1	(54.5)
Jolly Amylers	38.2	(245.4)	16.7	(107.2)	24.0	(159.7)	1.7	(10.9)	7.0	(4.53)	3.8	(24.4)
Bedford Yellow	35.1	(244.5)	18.3	(127.9)	16.5	(114.7)	1.3	(8.9)	3.4	(2.44)	12.4	(87.0)
Rubuste Nool	32.3	(221.1)	13.3	(91.0)	24.9	(169.9)	2.8	(18.8)	6.9	(4.74)	4.2	(28.7)
Lord Elco	35.9	(286.0)	15.5	(123.7)	20.6	(164.6)	1.9	(15.2)	6.2	(4.99)	3.8	(30.3)
White Eagle	26.4	(168.8)	7.5	(48.0)	29.2	(186.4)	2.1	(13.7)	6.2	(3.97)	3.2	(20.8)
Careless VT	28.5	(218.9)	11.6	(88.9)	21.6	(166.0)	1.4	(11.0)	3.6	(2.76)	0.7	(5.4)
Scotch Red Rough	8.6	(48.5)	4.9	(27.3)	26.1	(145.6)	4.5	(25.0)	12.0	(6.68)	11.7	(66.8)
Mitre	25.6	(182.0)	9.3	(66.1)	20.8	(147.2)	1.4	(10.2)	3.4	(2.43)	0.8	(5.6)
Cousen's Seedling	26.5	(212.1)	9.6	(76.1)	11.2	(89.4)	2.0	(16.0)	5.5	(4.46)	3.4	(27.9)
Careless (Kent)	27.1	(179.7)	10.3	(68.1)	14.3	(94.8)	1.5	(10.3)	3.9	(2.58)	0.7	(4.8)
Careless 3024	17.2	(129.0)	7.2	(54.2)	20.0	(150.1)	1.5	(11.2)	4.5	(3.38)	5.6	(41.8)
Lord Audley	19.3	(163.7)	6.1	(51.2)	15.0	(126.6)	1.3	(11.2)	3.4	(2.84)	8.8	(73.9)
Jubilee Careless	27.6	(186.2)	10.1	(68.0)	11.4	(76.3)	1.1	(7.5)	2.6	(1.78)	0.7	(4.8)
May Dulle	6.4	(45.7)	5.5	(39.5)	22.3	(158.8)	2.3	(16.1)	5.1	(3.64)	8.4	(69.1)
Victoria	14.5	(118.6)	6.9	(56.0)	17.0	(138.4)	1.0	(8.2)	3.8	(3.04)	5.1	(42.2)
Goutray's Earliest	12.3	(107.2)	5.3	(46.7)	17.8	(155.5)	1.1	(9.6)	3.4	(2.95)	6.7	(58.1)
Ajax	17.2	(116.5)	5.8	(39.4)	13.5	(91.7)	1.4	(9.6)	4.0	(2.70)	3.9	(26.5)
Mean	26.0	(177.2)	10.5	(71.9)	21.7	(146.6)	1.8	(12.2)	5.5	(3.68)	5.6	(38.5)
LSD	5.31	(24.87)	2.59	(12.36)	5.75	(29.94)	0.51	(3.63)	0.98	(0.532)	1.61	(11.50)

[a] The results are presented in descending order according to the sum of hydroxycinnamic acid derivatives for FW.

Table 3. Tentative identification of anthocyanins and hydroxycinnamic acid glycosides in gooseberry cvs with QToF/MS/MS.

Peak	Ret. time (min)	m/z	Fragments MS/MS	Tentative identification	Molecular formula
Anthocyanins					
3	7.13	[M+] 463.1303	301.0995	Peonidin-3-glucoside	$C_{22}H_{23}O_{11}$
3	7.27	[M+] 419.1190	287.055	Cyanidin-3-xyloside	$C_{20}H_{19}O_{10}$
4	7.42	[M+] 609.2129	301.0994	Peonidin-3-rutinoside	$C_{28}H_{33}O_{15}$
6	10.11	[M+] 595.1447	287.055	Cyanidin-glycoside	
7	10.36	[M+] 595.1480	(287.0814, 147.0570, 163.1326)	Cyanidin-coumaroyl-glucopyranoside	$C_{30}H_{27}O_{13}$
8	10.96	[M+] 609.1594	301.705	Peonidin-glycoside	$C_{28}H_{33}O_{15}$

Ret. time (min)	m/z	Fragments MS/MS	Tentative identification	Molecular formula
Hydroxycinnamic acid derivatives				
5.15, 6.10	[M − Na]+ 349.0893	(165.0545, 147.0570)	Coumaric acid hexoses	$C_{15}H_{18}O_8$
4.20	[M − Na]+ 365.0862	(181.0508, 163.0402, 145.0283, 135.0437, 117.0333, 89.0383)	Caffeic acid hexose	$C_{15}H_{18}O_9$
4.86, 6.42	[M − Na]+ 379.0999	(195.0651, 177.0546)	Ferulic acid hexoses	$C_{16}H_{20}O_9$
5.62	[M − Na]+ 409.1109	(225.076, 207.0652, 175.0391)	Sinapic acid hexose	$C_{17}H_{22}O_{10}$

3.1.3. Characterization and quantification of anthocyanins

A total of eight anthocyanins were detected in the pink and red gooseberry varieties. Cyanidin-3-rutinoside and cyanidin-3-glucoside were identified by comparison with authentic standards. The other anthocyanins were tentatively identified based on their accurate mass and the fragmentation pattern acquired with MS/MS in different voltages. (Table 3). The major daughter ion produced was the aglycon after the loss of the sugar moiety. Peak 3 corresponded to peonidin-3-glucoside, with small amounts of cyanidin-3-xyloside, also co-eluting under the same peak as revealed by the mass profile obtained with Q-ToF/MS/MS. Peak 4 was tentatively identified as peonidin-3-rutinoside. Peaks 6 and 7 had accurate mass at *m/z* 595.1447, and 595.1480, respectively, and after fragmentation they both yielded a major daughter ion with *m/z* 287.055 which was attributed to cyanidin aglycon. Peak **7** additionally had a daughter ion with *m/z* 147.0570 which is a characteristic fragment of *p*-coumaric acid. Jordheim et al. (2007), have characterised two isomers of cyanidin-3-*O*-*p*-coumaroyl-glucopyranoside in gooseberries by HPLC-DAD and NMR spectroscopy. Peak 7, furthermore had the same UV/Vis spectra (Fig. S1) as cyanidin-3-*O*-β-(6''-*E*-*p*-coumaroylglucopyranoside) as reported by Jordheim et al. (2007). Therefore it was tentatively identified as cyanidin-*p*-coumaroyl-glucopyranoside. Peak 8 exhibited a similar mass profile as peak 4 and it was tentatively identified as peonidin-3-glycoside. Peak 5 was present in very low amounts and it was not possible to acquire a clear LC/MS profile.

The anthocyanin concentrations in pigmented gooseberry cvs are presented in Table 5. Cyanidin-3-rutinoside (pigment 2) was the predominant anthocyanin identified in all the cvs followed by cyanidin-3-glucoside (pigment 1) and pigment 7. 'Scotch Red Rough' exhibited the highest total anthocyanin content with a sum of 265.8 mg kg^{-1} (FW), which is comparable with the levels of anthocyanins reported for other red gooseberry cvs and similar to the levels found in red currants (Djordjevic et al., 2010, Jordheim et al., 2007 and Määttä-Riihinen et al., 2004).

Table 4. Concentration of hydroxycinnamic acid glycosides (g kg^{-1}) in the seeds of 20 gooseberry cvs on a DW basis.[a]

Cultivar	Coumaroyl-hexose1	Coumaroyl-hexose2	Caffeoyl-hexose	Feruloyl-hexose1	Feruloyl-hexose2	Sinapoyl-hexose	ΣHCAD[b]
Heart of Oak	4.55	1.09	6.65	0.15	1.10	0.20	13.74
Guy seedling	2.26	0.64	4.19	0.12	0.64	0.19	8.04
Victoria	1.85	0.67	3.99	0.19	0.68	0.33	7.71
Careless VT	2.21	0.75	3.03	0.15	0.76	0.31	7.21
Nailer	1.77	0.64	3.33	0.12	0.65	0.33	6.84
Mitre	1.81	0.65	3.26	0.15	0.66	0.28	6.81
Jolly Amylers	1.70	0.57	3.23	0.12	0.58	0.37	6.57
Bedford Yellow	1.97	0.76	1.74	0.17	0.76	0.24	5.64
May Dulle	1.35	0.46	2.47	0.15	0.47	0.18	5.08
Ajax	0.91	0.35	2.89	0.12	0.36	0.30	4.93
Jubilee Careless	1.52	0.51	1.99	0.13	0.51	0.25	4.91
Careless 3024	1.29	0.50	1.95	0.14	0.50	0.31	4.69
Careless (Kent)	1.55	0.49	1.82	0.11	0.49	0.17	4.63
Cousen's Seedling	1.53	0.45	1.75	0.16	0.45	0.29	4.63
White Eagle	1.14	0.31	2.30	0.10	0.32	0.32	4.49
Goutray's Earliest	0.82	0.33	2.48	0.13	0.33	0.20	4.29
Rubuste Nool	0.98	0.28	1.81	0.08	0.28	0.25	3.68
Lord Elco	0.99	0.30	1.60	0.10	0.30	0.31	3.60
Lord Audley	0.71	0.24	1.75	0.11	0.24	0.15	3.20
Scotch Red Rough	0.61	0.21	1.72	0.16	0.21	0.05	2.96
Mean	1.58	0.51	2.70	0.13	0.51	0.25	5.68
LSD	0.356	0.15	0.917	0.036	0.151	0.077	1.393

[a] The results are presented in descending order according to the sum of hydroxycinnamic acid derivatives for DW.
[b] ΣHCAD = Sum of hydroxycinnamic acid derivatives.

3.2. Presence of non-structural carbohydrates

The concentration of total sugars on a FW basis, ranged between 31.4 g kg^{-1} to 80.2 g kg^{-1} (Table 6), which is consistent with previous reports (Mikulic-Petkovsek et al., 2012b and Viljakainen et al., 2002). Fructose and glucose were the predominant sugars in all cvs accounting for ~49% and ~40.5%, respectively, of total sugars. The mean ratio of sucrose was ~10% of the total sugars in gooseberry cvs. These results are in agreement with Giné Bordonaba and Terry (2008) who reported the same ratios of sugars in 17 UK-grown blackcurrant cvs, indicating a similar sugar profile of different species belonging to the *Grossulariaceae*. The ratio of fructose to glucose reported here varied from 0.9 to 1.3 and is within the range reported in red and blackcurrants (Milivojevic et al., 2012). The relative proportion of individual sugars in fruits can influence the taste properties such as sweetness (Milivojevic et al., 2012).

3.3. Total ascorbic acid

Ascorbic acid was determined as total ascorbic acid which comprised ascorbic acid and its oxidised form (dehydroascorbic acid). It was detected as a single peak at 3.1 min. The vitamin C content varied from 0.38 g kg^{-1} in 'White Eagle' to 0.85 g kg^{-1} (FW) in 'Ajax' (Table 7). These concentrations are up to 3-fold higher than previous reported for gooseberries (Häkkinen et al., 1999b, Pantelidis et al., 2007 and USDA, 2014), although they are similar to the concentrations reported in blackcurrants (Djordjevic et al., 2010), red currant, strawberry and lingonberry (Skrede et al., 2012) and within the range found in common berries (<1 g kg^{-1} FW) (Giné Bordonaba and Terry, 2011). The differences between these and earlier studies could reflect differences in sample preparation and analytical methods or more importantly differences among varieties.

3.4. Postharvest treatment of gooseberries

The two cvs selected for the storage trial in the following year, included the red cv 'Scotch Red Rough' which exhibited the most diverse profile of phenolics and had a high concentration of anthocyanins as well as the green cv 'Careless (Kent)' which is a popular commercial variety of gooseberry. For both cvs (per DW), the phenolic and sugar content at harvest point was compared between the two successive years. The composition of non-structural sugars between the two seasons was similar, with only sucrose in 'Careless Kent' dropping from 41.8 g kg^{-1} in the first year to 26.5 g kg^{-1} in the second ($p < 0.05$) (Table S1). Overall, the profile of phenolics was similar for both years; the most notable difference was the 2-fold drop in anthocyanin content in 'Scotch Red Rough' during the second year from 1477.8 mg kg^{-1} to 749.5 mg kg^{-1} (Table S1).

3.4.1. Biochemical and physiology changes during storage

In the present study the changes in major phenolic compounds, non-structural sugars, colour and ethylene emission were recorded for both cvs, in order to determine the biochemical and physiology changes in gooseberry fruit during air storage and the effect of external ethylene exposure. The water content was also monitored for both cvs throughout the storage period. 'Careless Kent' had a consistently higher water content compared to 'Scotch Red Rough' with average values at baseline of 87.7% and 81.8%, respectively. The results indicated a marginal water loss of ≤2% for both cvs, irrespective of treatment, which occurred after 7 days of storage (Table S2). Also there was no leakage recorded in any of the cvs.

3.4.2. Ethylene

The change in ethylene production rates for treated and control samples is presented inFig. 2. Ethylene emission remained at low levels throughout the storage period with a total mean of 5.8×10^{-3} ng kg^{-1} s^{-1}, which is significantly lower compared to other soft fruits such as blackberries and raspberries (0.035–0.35 ng kg^{-1} s^{-1} at 5 °C) (Mitcham et al., http:// postharvest. ucdavis.edu /PFfruits/ Bushberries/). An increasing trend was observed for the red cv with the control samples showing a sharp increase owards the middle of the storage period, but subsequently fell in the same levels as the treated samples. Ethylene emission levels remained relatively stable for the green variety throughout the storage period.

Table 5. Concentration of anthocyanins (mg kg^{-1}) in 20 gooseberry cvs on a FW and DW basis (values inside brackets). The results are presented in descending order according to the sum of anthocyanins per FW.[a]

Cultivar	1		2		3		4		5		6		7		8	
Scotch Red Rough	48.3^{d}	(267.5)	128.8^{h}	(714.6)	3.3^{c}	(18.5)	14.5^{c}	(82.9)	6.0b^{a}	(31.9)	5.7^{c}	(31.7)	57.3^{e}	(319.8)	1.9^{a}	(10.9)
May Dulle	33.6^{d}	(239.0)	36.3^{g}	(256.6)	2.0^{b}	(14.0)	2.9^{b}	(20.6)	1.5^{b}	(10.7)	2.6^{b}	(18.2)	21.1^{d}	(149.9)	0.4^{b}	(3.1)
Careless 3024	8.2^{ab}	(61.6)	19.5^{f}	(146.4)	t[b]		0.8^{a}	(6.2)	t		0.8^{a}	(5.7)	4.6^{c}	(34.4)	t	
Victoria	9.4^{b}	(83.9)	13.5^{c}	(109.7)	nd[c]		t		t		0.5^{a}	(4.4)	2.8^{a}	(23.1)	nd	
Goutray's Earliest	6.1^{ab}	(53.6)	8.3^{abc}	(72.0)	0.8^{a}	(6.9)	0.9a	(7.7)	0.6c	(5.0)	t		3.0^{a}	(26.1)	t	
Ajax	5.3^{ac}	(36.1)	9.3^{bc}	(63.1)	t		1.1a	(7.7)	t		t		2.1^{a}	(14.2)	nd	
Lord Audley	3.6^{c}	(30.5)	4.8^{a}	(40.9)	t		t		t		nd		1.4^{b}	(11.8)	nd	
Heart of Oak	1.0^{e}	(6.2)	5.4^{ab}	(31.7)	nd		t		nd		t		t		nd	
Bedford Yellow	t		1.8^{e}	(12.2)	t		t		t		t		t		t	
Nailer	t		1.2^{de}	(7.3)	nd		t		nd		nd		t		t	
Jolly Amylers	t		0.6^{d}	(4.3)	nd		t		nd		t		t		nd	
Mean	14.6	(9.73)	20.9	(132.6)	2.0	(13.2)	4.0	(25.0)	2.7	(15.9)	2.4	(15.0)	13.2	(82.8)	1.2	(7.0)
LSD					0.58	(4.05)	1.17	(2.75)	0.38	(2.76)	0.68	(4.22)			0.69	(4.59)

[a] The cultivars for which the levels of anthocyanins were below the detection limit (DL) or quantification limit (QL) have been omitted from the table.
[b] t = traces.
[c] nd = not detected. Means within the same column with no letters in common are significantly different. Data for which LSD is not provided have been log transformed and the significant differences are based on the analysis of variance (ANOVA) for the logarithmic data.

Table 6. Concentration of non-structural carbohydrates (g kg^{-1}) in 20 gooseberry cultivars on a FW and DW basis (values inside brackets).[a]

Cultivar	Fructose		Glucose		Sucrose		Total	
Guy seedling	37.5^{h}	(189.8)	31.7^{f}	(160.6)	10.9^{e}	(55.0)	80.2^{g}	(405.4)
Jubilee Careless	34.1^{gh}	(228.4)	29.0^{ef}	(193.5)	9.4^{de}	(61.4)	72.5^{fg}	(483.3)
Ajax	33.2^{fgh}	(225.2)	28.2^{def}	(191.7)	6.4^{cd}	(43.5)	67.8^{fg}	(460.3)
Scotch Red Rough	30.9^{fgh}	(173.9)	24.5^{cdef}	(138.2)	11.5^{e}	(63.4)	66.9^{fg}	(375.5)
Careless (Kent)	32.1^{fgh}	(212.9)	28.0^{def}	(185.9)	6.3^{cd}	(41.8)	66.4^{fg}	(440.6)
Rubuste Nool	32.5^{fgh}	(218.2)	27.4^{cdef}	(183.8)	6.0^{cd}	(44.0)	65.9^{fg}	(442.4)
White Eagle	30.8^{fgh}	(196.8)	27.7^{cdef}	(176.3)	6.6^{cd}	(41.9)	65.1^{fg}	(415.0)
Careless VT	31.6^{fgh}	(242.4)	27.3^{cdef}	(209.3)	4.5^{bc}	(34.8)	63.5^{fg}	(486.4)
Bedford Yellow	31.0^{fgh}	(218.2)	25.7^{cdef}	(180.7)	5.7^{c}	(40.0)	62.4^{fg}	(439.0)
Nailer	27.8^{efgh}	(166.1)	22.1^{bcde}	(132.1)	10.8^{e}	(64.0)	60.7^{f}	(362.2)
Heart of Oak	24.7^{cdef}	(143.5)	20.3^{bcd}	(118.1)	14.3^{e}	(82.5)	59.3^{f}	(344.2)
Cousen's Seedling	29.3^{efgh}	(235.4)	24.9^{cdef}	(200.2)	3.1^{ab}	(25.0)	57.4^{ef}	(460.7)
Mitre	27.3^{efg}	(195.0)	23.3^{cdef}	(166.4)	5.3^{c}	(37.7)	55.9^{def}	(399.1)
May Dulle	26.1^{defg}	(185.7)	22.0^{bcde}	(156.3)	4.6^{c}	(32.8)	52.7^{cdef}	(374.8)
Jolly Amylers	25.0^{bcdef}	(159.4)	20.7^{bc}	(131.9)	5.5^{c}	(35.2)	51.3^{bcdef}	(326.6)
Victoria	21.9^{abcde}	(178.8)	16.4^{ab}	(133.9)	2.9^{a}	(23.5)	41.2^{abcde}	(336.2)
Lord Elco	19.5^{abcd}	(155.7)	16.3^{ab}	(129.0)	2.9^{a}	(22.8)	38.6^{abcd}	(308.5)
Goutray's Earliest	19.3^{abc}	(168.5)	14.2^{a}	(123.4)	3.1^{a}	(27.2)	36.6^{abc}	(319.0)
Lord Audley	17.8^{ab}	(153.0)	14.0^{a}	(118.4)	2.8^{a}	(23.9)	34.6^{ab}	(292.6)
Careless 3024	16.6^{a}	(124.1)	12.0^{a}	(90.0)	2.8^{a}	(21.2)	31.4^{a}	(235.3)
Mean	27.5	(188.4)	22.8	(156.0)	6.3	(40.9)	56.5	(385.4)

[a] The results are presented in descending order according to total sugar content for FW. Means within the same column with no letters in common are significantly different. Results are based on the analysis of variance (ANOVA) for the log transformed data for all the sugars.

Table 7. Concentration of total ascorbic acid (g kg^{-1}) in 20 gooseberry cvs on a FW basis.[a]

Cultivar	Total ascorbic acid
Ajax	0.85
Careless VT	0.83
Jubilee Careless	0.81
Lord Elco	0.74
Careless (Kent)	0.74
Nailer	0.68
Scotch red rough	0.64
Heart of Oak	0.60
Victoria	0.56
Bedford yellow	0.55
Guy Seedling	0.54
Mitre	0.53
Cousen's Seedling	0.53
Careless 3024	0.52
May Dulle	0.51
Lord Audley	0.48
Robuste Nool	0.45
Goutrey Earliest	0.44
Jolly Amylers	0.43
White Eagle	0.38
Mean	0.59
LSD	0.049

[a] The results are presented in descending order according to total ascorbic acid content.

3.4.3. Phenolic compounds

The biochemical results, on a dry weight basis, showed that flavonol glycosides, which represent the major group of phenolic compounds in gooseberries, remained relatively stable for both treated and control samples throughout the storage period. 'Scotch Red Rough' had consistently higher concentrations of flavonol glycosides during the trial, as it is characteristic of the cv (Table S3).

Anthocyanins on the other hand exhibited an increasing trend in 'Scotch Red Rough' over time irrespective of treatment (Fig. 3). A marginal increase observed for the ethylene treated samples was not recorded towards the end of the storage period with both treated and control samples, reaching the same anthocyanin concentrations. This could indicate an initial positive effect of exogenous ethylene on the accumulation of anthocyanins which was counteracted by the sudden increase in ethylene production observed in the control samples towards the middle of the storage period. Other non-climacteric fruits, such as grapes, have exhibited a clear response to postharvest ethylene treatment with anthocyanins positively affected at the end of the treatment period (Becatti et al., 2010 and Becatti et al., 2014). These differences could be due to species differences and also differences in the treatment protocols applied. Grape berries were exposed to very high ethylene concentrations (1000 mg L^{-1}) and for a longer period of time (36 h), which could explain the acute metabolic responses to ethylene.

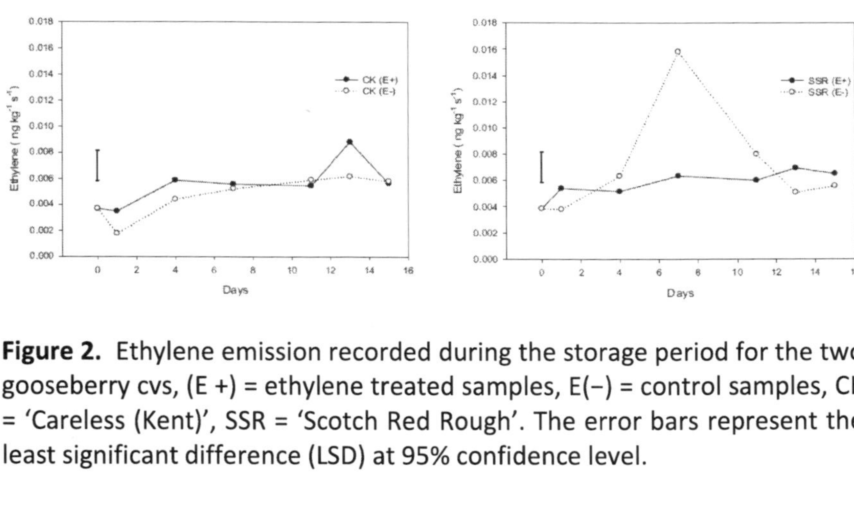

Figure 2. Ethylene emission recorded during the storage period for the two gooseberry cvs, (E +) = ethylene treated samples, E(−) = control samples, CK = 'Careless (Kent)', SSR = 'Scotch Red Rough'. The error bars represent the least significant difference (LSD) at 95% confidence level.

On the other hand, the accumulation of anthocyanins under air-storage, observed for the red gooseberry cv, seems to be a common trend among berry species, Kim et al. (2015), recently have demonstrated an increase in anthocyanin content in blackberries during storage at 1 °C, which was more pronounced if the berries were transferred to 20 °C towards the end of the storage period.

Earlier reports, suggest that gooseberry exposure to ethylene, can cause some yellowing/browning of the skin (Cantwell, 2002). In the current study, colour measurements revealed a decrease in L^* and C^* values over time, for both treated and control gooseberries (Table S2), which was not associated with any perceived colour deterioration. The decrease in L^* values could be attributed to several reasons including the observed accumulation of anthocyanin and perhaps gradual water loss. The change in C^* values, early in the storage period, may be related to an initial response to exogenous ethylene for the treated samples and the increase in ethylene emission for the control samples, which was further enhanced by water loss later in the storage period.

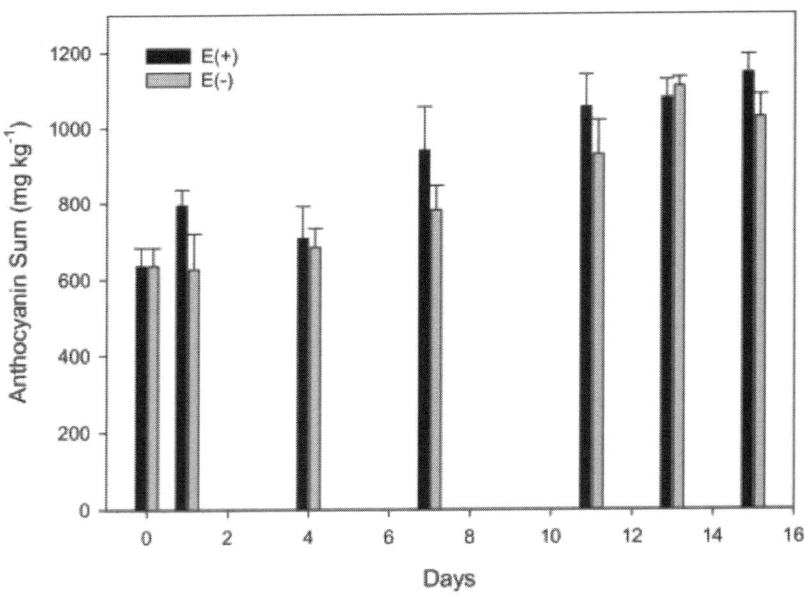

Figure 3. Concentration of total anthocyanins on a DW basis, in 'Scotch Red Rough' during the storage period. E(+) = Ethylene treated samples, E(−) = control. The error bars represent the standard error (SE) of the mean.

3.4.4. Non-structural carbohydrates

The concentrations of non-structural carbohydrates, for both cvs examined, remained relatively stable for both treated and control samples. The main difference in sugar composition between the two cvs was in the levels of sucrose, which were particularly elevated in 'Scotch Red Rough' throughout the storage period. The average concentration of total sugars for the two cvs, at baseline and during the storage period, is shown in Fig. 4. A previous study on 'Achilles' gooseberries stored under air and different CA conditions, showed that total soluble solids, remained stable for 7 weeks (Harb and Streif, 2004).

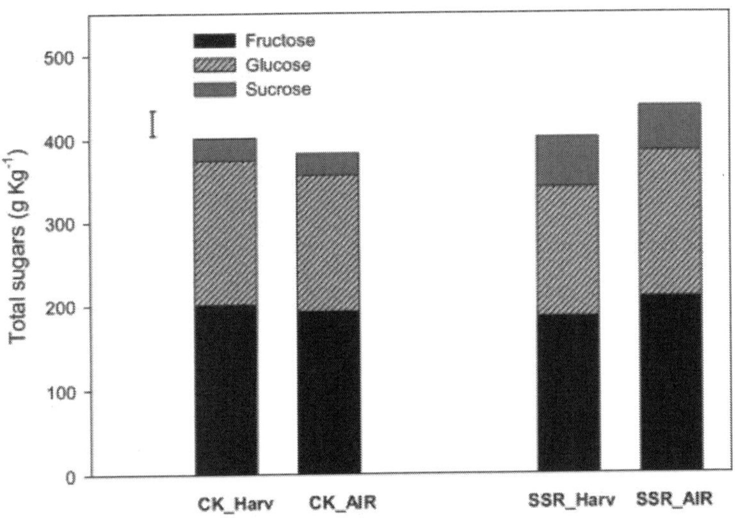

Figure 4. Average concentrations of non-structural carbohydrates on a DW basis, before and during storage for the two gooseberry cvs. CK_Harv = 'Careless (Kent)' at harvest, CK_AIR = 'Careless (Kent)' under air storage, SSR_Harv = 'Scotch Red Rough' at harvest, SSR_AIR = 'Scotch Red Rough' under air storage.

The biochemistry and physiology results overall, reveal that gooseberries did not show any clear response to the application of exogenous ethylene and are in broad agreement with previous reports classifying gooseberries as fruits with low sensitivity to ethylene (Cantwell, 2002).

4. CONCLUSIONS

The present study reveals the potential for the use of gooseberries as a rich source of bioactive compounds and demonstrates the distinguishing variability across cultivars and tissue types. Furthermore, it enhances our understanding of their postharvest behaviour especially in relation to their response to exogenous ethylene and the preservation of their nutritional quality under cold storage.

ACKNOWLEDGEMENTS

This study was supported by Unilever U.K. Central Resources Ltd, TSB 101125 (Innovate UK), and the Biotechnology and Biological Sciences Research Council (BBSRC). The authors are grateful to Dr M. Ordidge from the University of Reading for providing access to the National Fruit Collection and Farm Advisory Services Team (FAST) for helping with the sample collection.

REFERENCES

1. Bakowska-Barczak, A.M., Schieber, A., Kolodziejczyk, P., 2009. Characterization of Canadian Black Currant (Ribes nigrum L.) Seed Oils and Residues. J. Agric. Food Chem. 57, 11528–11536.

2. Barney, D.L., Hummer, K.E., 2005. Currants, Gooseberries, and Jostaberries. A Guide for Growers, Marketers, and Researchers in North America. The Haworth Press, Binghamton, NY, USA, pp. 266.

3. Becatti, E., Ranieri, A., Chkaiban, L., Tonutti, P., 2010. Ethylene and wine grape berries: metabolic responses following a short-term postharvest treatment. Acta Hortic. 223–228.

4. Becatti, E., Genova, G., Ranieri, A., Tonutti, P., 2014. Postharvest treatments with ethylene on Vitis vinifera (cv Sangiovese) grapes affect berry metabolism and wine composition. Food Chem. 159, 257–266.

5. Cantwell, M., 2002. Summary Table of Optimal Handling Conditions for Fresh Produce. In: Kader, A. (Ed.), Postharvest Technology of Horticultural Crops, 3. UCANR Publications, Oakland, California, USA, pp. 511–518.

6. Chiang, C., Kadouh, H., Zhou, K., 2013. Phenolic compounds and antioxidant properties of gooseberry as affected by in vitro digestion. LWT – Food Sci. Technol. 51, 417–422.

7. Dale, A., 2000. Potential for Ribes cultivation in North America. Hort. Technol. 10, 548–554.

8. Djordjevic, B., Šavikin, K., Zdunic, G., Jankovic, T., Vulic, T., Oparnica, C., Radivojevic, D., 2010. Biochemical properties of red currant varieties in relation to storage. Plant Foods Hum. Nutr. 65, 326–332.

9. Fait, A., Hanhineva, K., Beleggia, R., Dai, N., Rogachev, I., Nikiforova, V.J., Fernie Aharoni, A.R.A., 2008. Reconfiguration of the achene and receptacle metabolic networks during strawberry fruit development. Plant Phys. 148, 730–750.

10. Filipiak-Szok, A., Kurzawa, M., Szlyk, E., 2012. Determination of anti-oxidant capacity and content of phenols, phenolic acids, and flavonols in Indian and European gooseberry. Chem. Pap. 66, 259–268.

11. Folmer, F., Basavaraju, U., Jaspars, M., Hold, G., El-Omar, E., Dicato, M., Diederich, M., 2014. Anticancer effects of bioactive berry compounds. Phytochem. Rev. 295– 322.

12. Giné Bordonaba, J., Terry, L.A., 2008. Biochemical profiling and chemometric analysis of seventeen UK-grown black currant cultivars. J. Agric. Food Chem. 56, 7422–7430.

13. Giné Bordonaba, J., Terry, L.A., 2011. Ribes and Rubus [blackberry, currants and raspberry, etc.]. In: Terry, L.A. (Ed.), Health-Promoting Properties of Fruits and Vegetables, 1. Cabi, Oxford, UK, pp. 260–290.

14. Gopevac, D., Teševic, V., Vajs, V., Milosavljevic, S., Zdunic, G., Orpevic, B., Stankovic, M., 2012. Chemical composition of currant seed extracts and their protective effect on human lymphocytes DNA. J. Food Sci. 77, C779–C783.

15. Häkkinen, S., Heinonen, M., Kärenlampi, S., Mykkänen, H., Ruuskanen, J., Törrönen, R.,1999a. Screening of selected flavonoids and phenolic acids in 19 berries. Food Res. Int. 32, 345–353.

16. Häkkinen, S.H., Kärenlampi, S.O., Heinonen, I.M., Mykkänen, H.M., Törronen, A.R., 1999b. Content of the flavonols quercetin, myricetin,

and kaempferol in 25 edible berries. J. Agric. Food Chem. 47, 2274–2279.

17. Harb, J., Streif, J., 2004. Quality and consumer acceptability of gooseberry fruits (Ribes uva crispa) following CA and air storage. J. Hortic. Sci. Biotech. 79, 329– 334.

18. Hummer, K.E., Dale, A., 2010. Horticulture of Ribes. For. Path. 40, 251–263. Jordheim, M., Måge, F., Andersen, Ø.M., 2007. Anthocyanins in berries of Ribes including gooseberry cultivars with a high content of acylated pigments. J. Agric. Food Chem. 55, 5529–5535.

19. Kampuse, S., Muizniece-Brasava, S., Kruma, Z., Sabovics, M., Savri, S., Ruse, K., 2015. The influence of pretreatment technology and packaging material on the quality of fresh gooseberries during storage. Acta Hortic. 1071, 405–412.

20. Kim, M.J., Perkins-Veazie, P., Ma, G., Gina Fernandez, G., 2015. Shelf life and changes in phenolic compounds of organically grown blackberries during refrigerated storage. Postharvest Biol. Technol. 110, 257–263.

21. Lu, Y., Foo, L.Y., 2003. Polyphenolic constituents of blackcurrant seed residue. Food Chem. 80, 71–76.

22. Lu, Y., Foo, L.Y., Herbert Wong, H., 2002. Nigrumin-5-p-coumarate and nigrumin-5- ferulate, two unusual nitrile-containing metabolites from black currant (Ribes nigrum) seed. Phytochemistry 59, 465–468.

23. Määttä-Riihinen, K.R., Kamal-Eldin, A., Mattila, P.H., González-Paramás, A.M., Törrönen, R., 2004. Distribution and contents of phenolic compounds in eighteen scandinavian berry species. J. Agric. Food Chem. 52, 4477–4486.

24. Mattila, P., Hellström, J., Törrönen, R., 2006. Phenolic acids in berries, fruits, and beverages. J. Agric. Food Chem. 54, 7193–7199.

25. McKay, S., Van Eck, A., 2006. Red currants and gooseberries: extended season and marketing flexibility with controlled atmosphere storage. N. Y. Fruit Q. 14, 43– 45.

26. Mikulic-Petkovsek, M., Slatnar, A., Stampar, F., Veberic, R., 2012a. HPLC-MSn identification and quantification of flavonol glycosides in 28 wild and cultivated berry species. Food Chem. 135, 2138–2146.

27. Mikulic-Petkovsek, M., Schmitzer, V., Slatnar, A., Stampar, F., Veberic, R., 2012b. Composition of sugars, organic acids, and total phenolics in 25 wild or cultivated berry species. J. Food Sci. 77, C1064–C1070.

28. Milivojevic, J., Slatnar, A., Mikulic-Petkovsek, M., Stampar, F., Nikolic, M., Veberic, R., 2012. The influence of early yield on the accumulation of major taste and health-related compounds in black and red currant cultivars (Ribes spp.). J. Agric. Food Chem. 60, 2682–2691.

29. Mitcham, E., Crisosto, C.H., Kader, A.A., 2015. Bushberries: Blackberry, Blueberry, Cranberry, Raspberry: Recommendations for Maintaining Postharvest Quality. . (accessed 20-10-2015) http://postharvest.ucdavis.edu/ PFfruits/Bushberries/. Mitchell, C., Brennan, R.M., Cross, J.V., Johnson, S.N., 2011. Arthropod pests of currant and gooseberry crops in the U.K.: their biology, management and future prospects. Agric. For. Entomol. 13, 221–237.

30. Muizniece-Brasava, S., Dukalska, L., Kampuse, S., Murniece, I., Savri, S., DabinaBicka, I., 2015. The evaluation of physical parameters of gooseberries using different pretreatment technologies and packaging materials. Acta Hortic. 1071, 381–388.

31. Pantelidis, G.E., Vasilakakis, M., Manganaris, G.A., Diamantidis, G., 2007. Antioxidant capacity, phenol, anthocyanin and ascorbic acid contents in raspberries, blackberries, red currants, gooseberries and cornelian cherries. Food Chem.102, 777–783.

32. Planchon, V., Lateur, M., Dupont, P., Lognay, G., 2004. Ascorbic acid level of Belgian apple genetic resources. Sci. Hortic. 100, 51–61.

33. Pluta, S., Broniarek-Niemiec, A., Zurawicz, E., 2010. Productive value of eighteen gooseberry (Ribes grossularia L.) cultivars of different origin evaluated in central Poland. J. Fruit Ornam. Plant Res. 18, 197–209.

34. Russell, W.R., Labat, A., Scobbie, L., Duncan, G.J., Duthie, G.G., 2009. Phenolic acid content of fruits commonly consumed and locally produced in Scotland. Food Chem. 115, 100–104.

35. Skrede, G., Martinsen, B.K., Wold, A., Birkeland, S., Aaby, K., 2012. Variation in quality parameters between and within 14 Nordic tree fruit and berry species. Acta Agric. Scand. Sec. B Soil Plant Sci. 62, 193–208.

36. Terry, L.A., Chope, G.A., Giné Bordonaba, J., 2007. Effect of water deficit irrigation and inoculation with Botrytis cinerea on strawberry (Fragaria x ananassa) fruit quality. J. Agric. Food Chem. 55, 10812–10819.

37. Terry, L.A., Crisosto, C.H., Forney, C.F., 2009. Small fruit and berries. In: Yahia, E.M. (Ed.), Modified and Controlled Atmospheres for the Storage, Transportation, and Packaging of Horticultural Commodities. CRC Press, Boca Raton, USA, pp. 363– 395.

38. USDA, 2014. National Nutrient Database for Standard Reference Release 27. . (accessed 05-02-2015) http://ndb.nal.usda.gov/ndb/foods/show/2268? fg=Fruits+and+Fruit+Juices&man= & lfacet=& format = Abridged&count =&max=25 & offset=75& sort=&qlookup=.

39. Thompson, J.F., 2002. Transportation. In: Kader, A. (Ed.), Postharvest Technology of Horticultural Crops, 3. UCANR Publications, California, USA, pp. 259–269.

40. Viljakainen, S., Visti, A., Laakso, S., 2002. Concentrations of organic acids and soluble sugars in juices from nordic berries. Acta Agric. 52, 101–109.

41. Wang, L.-S., Stoner, G.D., 2008. Anthocyanins and their role in cancer prevention. Cancer Lett. 269, 281–290.

42. Wu, X., Gu, L., Prior, R.L., McKay, S., 2004. Characterization of anthocyanins and proanthocyanidins in some cultivars of Ribes, Aronia, and Sambucus and their antioxidant capacity. J. Agric. Food Chem. 52, 7846–7856.

Chapter 2

Post-harvest loss and quality deterioration of horticultural crops in Dire Dawa Region, Ethiopia

Mohammed Kasso[a], Afework Bekele[b]

[a]Department of Zoological Sciences, Addis Ababa University, P.O. Box 1176/Private P.O. Box 34387, Addis Ababa, Ethiopia

[b]Department of Zoological Sciences, Addis Ababa University, P.O. Box 1176, Addis Ababa, Ethiopia

ABSTRACT

The assessment on the causes of post-harvest loss and quality deterioration of horticultural crops was carried out in Dire Dawa Administration from March 2011 to December 2012 in the eleven randomly selected representatives Peasant Associations. Stratified and multi stage random sampling techniques were used to sample representative Peasant Associations and respondents (n = 296). Both qualitative and quantitative data were gathered through questionnaire, focus group discussions, interview and observations. Data were analyzed using appropriate descriptive statistics. Climate and weather conditions, harvesting and handling techniques, packaging, storage and transportation facility, market situation, dust from cement factory, disease and pest animals were recorded as major causes for post-harvest loss. The severe post-harvest loss and quality deterioration of horticultural crops mainly occurred during harvesting followed by marketing, transporting and storage. Poor quality equipment and materials usage caused tremendous mechanical, physiological and pathological damages on horticultural crops. To minimize losses, different traditional methods were practiced by the local community. The highest post-harvest loss was recorded for tomato (45.32%) followed by mango (43.53%), whereas the least post-harvest loss was recorded for coffee (15.75%). Post-harvest loss ranging from 20% to 50% was recorded in between marketing and consumption. This can be used as a good indication as all concerned bodies should aim for development of effective and efficient policies and strategies to solve existing problems.

KEYWORDS

Agricultural products; Dire Dawa; Horticultural crops; Post-harvest loss; Quality deterioration

1. INTRODUCTION

Agriculture is the mainstay of the Ethiopian economy (Bezabih and Hadera, 2007). The country has highly diversified agro-ecological conditions for the production of different cereal, legumes and horticultural crops (Bekele, 1989 and Milaku, 2005). More than 85% of its population is engaged in small scale agricultural production as a major means of livelihood (Bekele, 1989, Fantahun and Williamson, 2001 and Bezabih and Hadera, 2007). It contributes for about 50% of the country's gross domestic product (GDP), over 90% of export earnings and 73% supplies of raw materials for agro-based domestic industries (Milaku, 2005 and Bezabih and Hadera, 2007).

Alazar (2007) defined horticultural crops as crops, which are not staple cereals or major industrial crops, but are mainly eaten for their contribution to the flavor and interest of food with flexible consumption based on price, quality and supply. This includes fruits, vegetables, flowers, spices, medicinal and aromatic plants and plantation crops (Choudhury, 2006).

Horticultural crop production is the major farming system in the eastern part of Ethiopia such as in Dire Dawa Administration, Fedis, Haramaya, Kombolcha, Kersa, Meta, Kurfa Chelle, Grawa, Jarso and Gemechis.

Particularly in the areas where water is available and farmers have access to the market, horticultural production is used as a major source of cash income for households (Milaku, 2005 and Bezabih and Hadera, 2007).

Although human and material resources are devoted for planting, irrigation, fertilizer application, 50% of horticultural crops are lost due to post-harvest loss (Saxena et al., 1990, Alazar, 2007 and Olayemi et al., 2010). Post-harvest losses and quality deterioration of horticultural crops are mostly caused by pests, microbial infection, natural ripening processes and environmental conditions such as heat, drought and improper post-harvest handling (Idah et al., 2007 and Olayemi et al., 2010). It occurs through all or at least one of post-harvest activities such as harvesting, handling, storing, processing, packaging, transporting and marketing (Mrema and Rolle, 2002).

Even though only limited studies were conducted in Ethiopia, post-harvest loss and quality deterioration of horticultural crops are estimated to be 25–40% (Yohannes, 1989). Particularly, in the eastern part of the country, limited post-harvest improvement studies have been carried out for locally consumed fruits and vegetables. However, fruits such as banana, orange, lemon, pineapple and avocado which exported to Arab countries, Europe and the Middle East are relatively graded and packaged appropriately (Alazar, 2007).

Reduction of post-harvest losses and quality deterioration are essential in increasing food availability from the existing production. Minimizing this loss has a great significance for food security, economic growth and welfare of the society. Therefore, this survey was conducted to assess the principal causes of post-harvest loss and quality deterioration of horticultural crops in Dire Dawa Administration and to suggest solution to the problems.

2. MATERIALS AND METHODS

2.1. Study Area

This study was carried out in Dire Dawa Administration located in the eastern part of Ethiopia. The region is administratively divided into nine urban and 38 rural peasant association clustered into four clusters, which covers 98.7% of the total area. According to the report of CSA (2008), the total population of the Dire Dawa Administration is estimated to be 341,834.

Dire Dawa Administration is characterized by three broad traditional Agro-ecological zones mainly based on altitude, moisture and physio-geography; namely *"Kolla"* below 1500 m asl and *"Woina Dega"* (above 1500 m asl) and *"Dega"* (above 2000 m asl). The topography of the area constitutes from very steep high mountains to flat plains, where its altitude ranges from 950 to 2260 m asl. Ecologically, the Administration covers desert and semi-desert scrub and shrubland ecosystems. It is characterized by arid and semi-arid climate with low and erratic bimodal rainfall with about 677 mm and high monthly mean maximum temperature that ranges from 28.1 °C to 34.6 °C, and monthly mean minimum temperature ranges from 14.5 °C to 21.6 °C.

Dire Dawa Administration has favorable and ideal climatic condition as most horticultural crops require high temperature. It has high potential for horticultural crops for earning cash and as source of food for most producers. Horticultural crops such as vegetables (onion, tomato, pepper and cabbages), cash crops (khat, coffee and ground nut) and fruit crops (papaya, mango, orange, mandarin, sugar apple, custard apple, date palm, banana and guava) and tuber crops (potatoes, sweet potatoes and onions) were cultivated in more than 7300 ha of land in different parts of the administration (CSA, 2008).

As Dire Dawa is one of the commercial towns of Ethiopia, the administration serves transportation of many of horticultural products such as khat, vegetables, fruits and tuber crops for domestic use and for export. It serves as a medium for export of horticultural crops to Somalia and Djibouti (Bezabih and Hadera, 2007). The Kafira market, the old and historical market, is used as a terminal market for vegetables and fruits sold by small-scale traders coming from its surroundings.

2.2. Methods

The present study was carried out from April 2012 to December 2012. Within the study sessions, the major factors for post-harvest losses and quality deterioration of horticultural crops produced were assessed in the selected representative rural peasant associations. The extents of post-harvest loss and quality deterioration were estimated from the sample analysis and respondents' estimation. The selection of representative villages and householders or respondents from each stratum was based on a multistage systematic random sampling technique. Representative villages were selected first, and then the representative habitats and respondents were selected from each village.

2.2.1. Preliminary Survey

A preliminary survey was conducted during the first month of the study to identify the main rural Peasant Association that produces fruits and vegetables from irrigated fields. During this survey, all available and relevant information about the area was gathered. As producers are the main actors and risk takers for post-harvest loss and quality deterioration of horticultural crops, most of the intensive study was conducted in rural Peasant Associations. Based on the observation types of horticultural crops produced in the Administration and other relevant information, 11 Peasant Associations were randomly selected with consideration of the inclusion of different parts and agro-climatic zones of the Administration. The commodities that decided to be studied as horticultural crops were vegetables, fruits, tuber crops and cash crops.

2.2.2. Sampling Design

A two stage random sampling technique was employed to select respondents from each of the selected Peasant Associations. At first, representative Peasant Associations were selected. Then from the selected Peasant Associations, a total of 296 representative respondents were selected based on the quota or proportionate system. In addition to producers, respondents were also purposively selected from wholesalers, retailers and consumers from Dire Dawa town and from selected representative Peasant Associations.

2.2.3. Data Collection

A cross sectional type of data was collected on the principal causes of post-harvest loss and level of damage on horticultural crops with the help of pretested, structured, open and close-ended questionnaires, focus group discussion and harvesting and marketing observations and secondary data sources or document analysis.

As the result of the preliminary study indicated that rural communities participate on horticultural crop production and suffer more from post-harvest loss than urban community, more number of respondents were selected from rural Peasant Associations or from horticultural crops producers.

Primarily with the help of pretested structured open and close ended questionnaires information on types of post-harvest loss, level of post-harvest loss, mode of transport, packaging materials used, market related issues and attitude of respondents on post-harvest loss of fruits and vegetables was collected. In addition to the producers or farmers, data were also collected from wholesalers, retailers and consumers. Besides the primary data collection, guiding questions and checklist were also developed to gather general and specific information from producers, wholesalers, retailers and consumers. From focus group discussions and key informant interviews, information about the dominant horticultural crops produced in the area, agricultural practice related to post-harvest loss, means of harvesting, storage and transportation of horticultural crops, factors of post-harvest loss, and extents of post-harvest and marketing information was collected. Focus group discussions were held with community leaders, key informants, elders, youth and women farmers and responsible persons of different institutions and knowledgeable people on the subject in the study areas covering the entire selected representative Peasant Associations.

The estimation of post-harvest loss and quality deterioration on horticultural crops at different stages from production to consumption was measured by different methods such as estimating losses at producers, wholesalers, retailers and consumers level by calculating differences between fresh received and damaged products. Data were also collected on the type of items affected, the periods, stage, place and parts in which damages occurred. The post-harvest loss and quality deterioration at producer level was quantified based on the perceived estimated percentage losses and quality deterioration from the total harvested quantity. The post-harvest loss and quality deterioration of horticultural crops at collectors, wholesalers and retailers was estimated as the difference between quantity purchased and quantity sold in relation to total quantity purchased. The marketable quality of fruits was subjectively assessed following Awole et al. (2011)procedures for visual parameters of size, color, firmness, surface defects, sign of mold growth and shrinkage.

2.2.4. Data Analysis

The collected data were tabulated and organized in graphs, charts and tables and analyzed using descriptive statistics using SPSS version 16.0 Statistical Software. Appropriate statistical methods such as frequency, average, standard deviation, and percentage were used. The percentage of post-harvest losses and quality deterioration at different channels of post-harvest was measured by using the modified equation suggested by Debele et al. (2007):

Post-harvest loss and quality deterioration (%)

$$= \frac{W1 - W2}{W1} \times 100$$

where:

$W1$ = the original weight (kg) of given horticultural crops.

$W2$ = weight (kg) of given commodity after periodical interval of storage and transportation time appropriate to be sold or consumed.

Producers income loss due to middle men (brokers) was computed in percentage:

Producers income loss (%)

$$= \frac{P2 - P1}{P2} \times 100$$

where:

$P1$ = price of horticulture commodity at farm gate.

$P2$ = price of horticulture commodity at final accessible market (Dire Dawa).

3. RESULTS

3.1. Factors for Post-harvest Loss

As indicated in Table 1, climate and weather conditions, harvesting and handling techniques, packaging, storage and transportation and market situations were mentioned by almost equal proportion of respondents. Factors such as disease and pest animals were mentioned by relatively less number of respondents. High temperature and low humidity and other climatic factors such as wind and flood affected the production of horticultural crops.

Table 1. Major factors for post-harvest loss.

Factors	Frequency	Percent
Climate and weather conditions	58	19.6
Harvesting and handling techniques	58	19.6
Packaging, storage and transportation facility	57	19.3
Market situation	53	17.9
Disease	37	12.5
Pests	33	11.1
Total	296	100

Post-harvest loss and quality deterioration during harvesting and handling techniques was identified by 58 (19.6%) of the respondents (Table 1). They were not aware of contamination of microorganisms and pollution with heavy metals and pesticides associated with harvesting and handling. They did not use the recommended and suitable equipment and materials for harvesting and handling. These aggravated the risks of post-harvest loss and quality deterioration of horticultural crops.

Post-harvest loss and quality deterioration of horticultural crops occurred due to lack of proper care, use of inappropriate harvesting equipment and materials and lack of motivation and interest to improve and upgrade the harvesting and handling techniques from time to time.

Problems related to packaging, storage and transportation facilities as factors for horticultural crops loss and quality deterioration were stated by 57 (19.3%) of the respondents.

Marketing situation as major cause of post-harvest loss and quality deterioration was listed by more than 17% of the respondents. The distance of the market, low price, lack of proper means of transportation and conditions of road were mentioned as discouraging factors. In addition, producing similar type of horticultural crops at the same time has affected the market value. Fungi and bacteria were mentioned as the major cause of post-harvest loss and quality deterioration by 37 (12.5%) of the respondents during marketing (Table 1).

Pest animals such as arthropod, nematodes, birds and mammals were cited by 11% of the respondents as the major causes for post-harvest loss and quality deterioration of horticultural crops. Mechanical damages such as bruising, peeling, skin breakage, egg laying and larvae development inside the tissue were ascertained as the major damages resulted from the pest animals.

Pre-harvest loss of horticultural crops in particular on fruits and vegetable was due to the environmental impact of cement factory. Although the problem was not common among the entire selected representative Peasant Associations, the impact from Pioneer Cement Factory was high in the Genda Rigie Peasant Association. Field samples taken from the farmer Yahaya Haji Wadi indicated the annual estimated production loss of 28,160 kg of mango, 500,000 kg of orange, 4000 kg of tamarind and 5000 kg of custard apple based on the average production before the damage. The areas near the cement factory face problems of waste water disposal into the farm. As the area had shortage of water, farmers irrigate their farm. Furthermore, the smoke, dust and heat released from the factory increase the temperature and cause eye burn becoming obstacle to work in the farm. Dust deposited on fruit bearing trees has highly affected harvesting of fruits.

More than 57% of the respondents harvest their crops in the morning to transport to the market for sale. Harvest of most horticultural crops was during the coolest time of the day like in the early morning or evening.

About half of respondents harvest during the coolest time of the day and about 45% of respondents prefer medium temperature for harvesting the horticultural crops. The majority (51.7%) of the respondents preferred to harvest at the time when the humidity was moderate. However, comparatively few respondents preferred to harvest at any time and at high relative humidity.

High horticultural crop post-harvest loss and quality deterioration occur during harvesting, marketing, transporting and storage. The popularly mentioned time of horticultural crops loss and quality deterioration was during harvesting followed by marketing.

Most of the materials and equipment used were poor in quality. For harvesting they use hand picking or collecting with equipment such as sickle, spade, hoe, axe, ladder locally named '*baalla*', cloth, basket, canvas, sack, plastic bucket, palm oil container, aluminum or plastic bowel, oil can and thatched and weaved baskets and containers.

The packaging materials and equipment used were mostly rudimentary type. Baskets, jute sacs, '*tubo*', wooden box, plastic crates, leaves and branches of different plants, aluminum and plastic bowel, oil can, used cloth, cloth bags and rope were the most common packing materials and equipment.

Although plastic crates were well designed to prevent physical damage and is easy to handle as packaging containers, it was not used by most respondents due to its limited distribution. It was only used for relatively delicate horticultural items. They use broken and dirty crates. To overcome the limited distribution of plastic crates, the respondents share to use it in common. Jute sacks were widely used for almost all types of horticultural crops primarily due to its least price and easy availability and strength.

Several means were used to transport the horticultural crops from farm to temporary storage or to the market. These include the human labor, donkey, mule and horse cart, camel, bajaj, minibus locally called '*Daina*', mini and Isuzu trucks. The limited availability of the means of transport, long and rough road, climatic condition of the area and the rudimentary packing materials increases the post-harvest loss and quality deterioration of the horticultural crops.

The diverse types of containers and equipment used for packaging also served as a measuring unit during marketing. Most of the measuring units or containers used were not standardized and uniform size but rather it was based on conventional type. The measuring units or containers used for marketing were wooden box, plastic crates, aluminum and plastic bowel, jute sack, *'tubo'*, *'safedi'* and bundle. The measuring units were categorized according to their size, which also serve as pricing unit in the marketing of the products.

Although packing, transport and storage are known to be used to mitigate the post-harvest loss and quality deterioration by minimizing a gap between producer and consumer and between harvesting and consumption, the producers did not have suitable storage facilities and marketing sites. They lack proper marketing site in Dire Dawa town or Kafira, and forced to use the flooded area, which was filled with municipal waste and filth. In addition, their products were also exposed to direct sunlight until they were collected by the end users with a least price.

The temporary storage and warehouses constructed in different clusters and Dire Dawa Fruits and Vegetables market center buildings were not completed and started to give service. The planned buildings were not completed and fulfilled with necessary facilities. Even its construction has been interrupted.

As indicated in Table 2, the degree of the level of awareness of the respondents on the usage and practice of packaging materials were different. Although for perishable and delicate horticultural products, the packing materials used must have property of protection of the commodity from compression, collapse and mechanical damage, about 44% of respondents did not use such materials.

Except for sensitive items such as khat, lettuce, spinach and cabbage, the majority of respondents do not use vapor barriers. However, 46.6% of the respondents used to select the packaging materials that provide appropriate vapor barrier, even though not observed using.

Simple mechanical packing systems for volume-fill or tight-fill were known to be used by more than 72% of the respondents in different containers to reduce the vibration (Table 2). However, more than 27% of the respondents did not apply and practice this method.

As observed from Table 2, more than 62% of respondents agree on the practice of the selection of packaging materials based on the type and quality of the crop. However, majority of the producers did not sort and pack their horticultural products according to type and quality of the horticultural crops.

Table 2. Usage and practice in selection of packaging materials by the respondents.

Practices	Usages (n = 296)	
	Yes (%)	No (%)
Using non-collapsible and vented packages	55.7	44.3
Packaging containers appropriately filled	66.9	33.1
Packing materials with appropriate vapor barriers	46.6	53.4
Use of simple mechanical packing systems for volume-fill or tight-fill	72.3	27.7
Sorting and packing according to type and quality	62.2	37.8

Different major selection criteria were used by producers for the choice of packing materials. Nearly more than 41% of the respondents select the packing materials based on protection ability against temperature fluctuation, moisture loss and microorganism and other waste contamination. However, to overcome and minimize the risks associated with inappropriate packaging materials, they use the locally available plastic bags, wooden boxes with liner, metal cans and plastic cans for moisture loss sensitive horticultural crops. For items that require ventilation and air circulation, they perforate plastic bags, jute sacks or liners at regular distance to allow air circulation and prevent condensation. The rapping of horticultural crops with locally available branches, leaves, grasses, herbs and used cloths was also used for moisture retention and to reduce the possibility of cross-transfer of odors and disease.

More than 53% of the respondents protect post-harvest loss and quality deterioration of their horticultural crops by using different traditional or cultural methods. Although about 30% of the respondents use chemical or modern methods to protect post-harvest loss and quality deterioration, in reality they use it mostly during pre-harvesting agricultural operation. The rest use both traditional and chemical methods for protection.

The post-harvest loss and quality deterioration were prevented by traditional methods such as tree shade, storing in small hut or house, covering with clean straw, leaves and sorghum stalks for insulation and to avoid direct sun burn.

Another traditional method used to protect post-harvest loss and quality deterioration of horticultural crop was maintaining relative humidity by reducing temperature of the product or container or storage in order to reduce the evapotranspiration. The second technique used to maintain the relative humidity was by increasing the moisture of the air around the commodity by sprinkling, spraying and wetting the floor of store room, container or the commodity itself. In addition, vapor barrier materials such as polyethylene liner and bags and sacks, cloth coated boxes, fresh leaves and grasses and other variety of inexpensive and recyclable packaging materials were used. Despite these, wilting, shriveling, shrinking, and flaccid, decrease in weight and textural changes were commonly encountered during the study.

Curing of tuber crops such as sweet potatoes and potatoes was practiced to be stored for a length of time. They store the product at high temperature and high relative humidity in order to minimize wilting and decaying.

More than half of respondents sell their products after harvesting in the field or by taking it to accessible marketing sites with the exception of some commodities, which need to be stored to mature or for relatively less perishable products. Most of the time, resident houses were frequently used for temporary storage (43%). The use of farm field for temporary storage was practiced by a few respondents for some commodities by storing under big tree, caves or burying under the soil. The temporary storage of horticultural crops in the farm and human residence was in poor quality.

Major reasons and type of losses with their estimated post-harvest loss and quality deterioration of the commonly cultivated horticultural crops are shown in Table 3. Post-harvest loss of 45.32% for tomatoes was recorded followed by 43.35% for mango. The least post-harvest loss was recorded for coffee. Producers in many sites revealed the occurrence of up to 100%, particularly for tomatoes and onion due to decay from lack of transportation and market.

Nearly 60% of the producers sell their horticultural crops in Dire Dawa town, whereas the rest sell on farm field to rural collectors or brokers or in the nearby towns and villages. About 42% of the respondents sell their horticultural products to the retailers or to the rural collectors who act as brokers, whereas 38.2% sell directly to wholesalers, 15.2% to consumers in the Dire Dawa town or nearby suburban or rural community consumers. Although most of the times, they tend to share with their neighbor, more than 4% of the respondents reported to sell some horticultural crops to their neighbor consumers who did not produce that particular product.

Table 3. Type and estimated post-harvest losses and quality deterioration of the dominantly cultivated horticultural crops.

Horticulture	Types of major loss	Estimated loss (%)
Tomato	Abrasion, bruise, rupture, softening, shrivel, sore, bleach, squash, crash, over ripen	45.32
Mango	Wound, scratch, rotting, bleach, squash, puncture	43.53
Potatoes	Scratch, flaccid, decay, bleach, spot, compression, wound, wilting, crush, sprouting	37.15
Orange	Abrasion, discolor, sour, decomposition, shrinkage, rough and thick peal	35.58
Mandarin	Abrasion, discolor, sour, decomposition, shrinkage, rough and thick peal	34.25
Papaya	Scratch, flaccid, decay, bleach, spot, compression, lesion, crash	30.31
Khat	Wilting, leaf shot burn, weight loss, leaf puncture	27.34
Onion	Decay, wilt, shrink, flaccid, sprouting	25.21
Guava	Skin scratch, bruising, rotting, softening, discolor	23.10
Green pepper	Flaccid, decay, wilting, color change	22.54
Banana	Decay, softening, bruising, chilling, peal split and breakage, skin graze	19.87
Coffee	Shrivel, bleach, shrunk seed	15.75

The majority of the respondents believe road and transportation had direct impact on the post-harvest loss and quality deterioration of their horticultural crops. Particularly for Kulayu and Debele Peasant Association producers, the loss was more than half and in some occasion all of their produce, due to problems related to transportation and road. Although almost all peasant associations were connected by either temporary or permanent gravel or earthed road, appropriate vehicle for the transportation of their horticultural crops was lacking. Many of the vehicles used for transportation in the rural peasant associations of the Administrations were also old and not properly repaired. These problems have contributed a lot for the loss and quality deterioration of horticultural crops.

Almost half of the respondents did not carry out market assessment. They tried to sell their products randomly. However, 49% of the respondents carry out market assessment particularly for coffee and for other relatively less perishable products. They try to sell their products following the updated and appropriate price in farm gate or other market places.

Most of them believed they do not get better price for their products due to different reasons. For most horticultural crops, it was observed that producers lose more than 50% of the cost when sold at farm gate. The other reason was damage and loss of quality during transportation and other pre- and post-harvest activities.

3.2. Post-harvest Loss and Quality Deterioration During Trading and Consumption

The major factors for post-harvest loss and quality deterioration of horticultural crops during marketing were due to problems related to mishandling in packaging, transportation, storage and display for sale.

Most of the packing materials and containers used were of poor quality and unfit and inappropriate for the products that lead to mechanical damages such as bruising, abrasion, sprouting, decay, chilling, rotting and softening. Tomatoes were packed in collapsible plastic containers that cause great mechanical damage. The use of public transportation for transporting passengers and commodities together, overloading and stacking and high temperature, rough road with high vibration and collision, lack of vehicle and its high cost and the use of draft animals and humans for transporting long distances in the absence of proper packing and stacking were major problems. They do no sort out ripe and unripe, ethylene producers from ethylene sensitive products, odor producers from odor sensitive products. Absence of warehouses for horticultural crops and use of human residence or working room as store were also factors for deterioration of products.

Export of horticultural crops to Djibouti and Somalia was not as such standardized. For transportation, they use none-refrigerated and none-ventilated vehicles such as ISUZU, and Rail Trucker. As a compensation mechanism of lack of refrigerated vehicles, transportation was during the night.

Most of the post-harvest loss and quality deterioration of horticultural crops at consumer level occur due to lack of proper handling and subsequent damages during harvesting and marketing process. However, damages also occur during storage, packing and transportation. Most consumers did not use proper storage for different commodities. As a result, the loss for horticultural crops was 20–50%, and for cabbage and tomatoes it was >50%.

4. DISCUSSION

The horticultural crops are inherently liable to deteriorate under different climatic and other circumstances due to their high moisture content (Kitinoja and Kader, 2002). Moreover, as they are biologically active and carry out transpiration, respiration, ripening and other biochemical activities, they tend to loss and deteriorate through time. This makes the post-harvest losses to occur in the field, packing areas, in storage, during transportation and marketing. Severe losses occur because of environmental conditions, poor facilities, lack of know-how, poor management, weak marketing processes or simply due to carelessness of farmers. Proper storage conditions, temperature and humidity are needed to lengthen the storage life and maintain quality of horticultural crops (Kitinoja and Kader, 2002). Dire Dawa Administration is characterized by high temperature combined with low relative humidity (BFED, 2010). This has a negative effect on most horticultural products as a major factor for post-harvest loss and quality deterioration.

Horticultural crops may be characterized as being either climacteric or non-climacteric, depending on their respiratory pattern. Climacteric fruits can be harvested when mature but before the onset of ripening. After the climacteric, the respiration rate slows down as the fruit ripens and develops quality (Sirivatanapa, 2006). Thus, in the present study the harvesting of mandarin without proper maturity will lead to post-harvest loss and quality deterioration.

Fresh produce needs low temperature and high relative humidity during storage and transportation (Choudhury, 2006). Therefore, reducing the temperature and increasing the relative humidity are the primary means of maintaining product quality during storage and transportation. Reduced temperature decreases the physiological, biochemical and microbiological activities, which causes quality deterioration (Thompson et al., 1998).

Horticultural crops can be contaminated with different microbial and chemical contaminants during handling and processing and become source for infectious microorganisms (Moy, 2005). Mechanical damage during harvest can become a serious problem by disposing it to decay, increasing water loss, respiration and ethylene production rates, which leads to deterioration (Kitinoja and Kader, 2002). Bruises and other mechanical damage affect appearance and also provide entrance to decaying organisms (Olayemi et al., 2010).

From the three cement factories operating in the Administration, the Pioneer Cement Factory was blamed to cause a great loss on the nearby horticultural crops in Genda Rigie Peasant Association. Previous studies by Asubiojo et al., 1991 and Saralabai and Vivekanandan, 1992 revealed that the cement dust deposition in large quantities around cement factories causes changes on the growth and biochemical characteristics of crops and on the physical and chemical properties of the soil. The areas near the cement factory also face problems of alkalization due to high deposition of alkaline cement dust and ash. Its high dust deposition on vegetation reduces the productivity and concentration of chlorophyll (Liu et al., 1997 and Raajasubramanian et al., 2011). Cement factories are sources of pollutant to surrounding areas through dust deposition that affect photosynthesis, respiration, stomata functioning and productivity. It also increases the mortality of young branches leading to a reduction in biomass and number of fruits that lead to more than half of economic yield loss (Singh and Rao, 1980, Abdel-Rahman, 2012 and Abdel-Rahman and Ibrahim, 2012). The dust falling on the leaves also affects photosynthesis, stomata functioning and productivity. The impact is great around 500 m radius and slightly reduced at 5 km away from the cement factory (Abdel-Rahman and Ibrahim, 2012). Fakhry and Migahid (2011) concluded in their study that the influence of cement factories in arid regions is tremendous in vegetation diversity and the responses of individual species and the chemistry of the soil. Thus, if the current effects are not regulated, all horticultural crops and other biodiversity in the potential distance range will be vanished in the near future.

The majority of respondents seem to harvest during relatively appropriate time concerned with temperature and humidity, although night or early morning harvesting is used to lower internal temperatures and used for reduction of the energy needed for subsequent cooling (Kitinoja and Kader, 2002).

The severe horticultural crop post-harvest loss and quality deterioration were recorded mostly during harvesting followed by marketing, transporting, storage and in some cases through the entire channel. This is because fresh produce after harvest continues the process of respiration and transpiration until its reserved food and water are exhausted (Sirivatanapa, 2006). This physiological process is influenced by temperature, composition of surrounding air, and humidity of environment.

Although harvesting was carried out by hand rather than machine, some horticultural crops can be severely damaged by careless and inappropriate harvesting Absence of the use of maturity index as a standard cause high post-harvest loss and quality deterioration (Kitinoja and Kader, 2002). The containers used by pickers should be clean, smooth inside surfaces and free from rough edges (FAO, 1989). In addition, harvesting practices cause little mechanical damage by gentle digging, picking and handling (Kitinoja and Kader, 2002). Pickers should be trained to empty bags or baskets with care (Kitinoja and Kader, 2002). However, the producers throw mango, guava, mandarin, orange and papaya on to the ground from more than 3 meters high fruiting trees, that cause mechanical damages.

The use of transportation on rough road and open and closed lorry including public buses causes mechanical damage to horticultural crops as a result of vibrations and high temperature (Singh and Singh, 1992). The use of poor packaging material that restricts ventilation will also cause post-harvest loss and quality deterioration of horticultural crops (Olayemi et al., 2010).

The desired level of development in horticulture has not been achieved because of a number of constraints. Due to absence of proper storage and marketing facilities, and seasonal surplus, farmers are forced to sell their products at thrown-away prices and at the earliest opportunity after harvest (Wei et al., 2001). This also forces producers to sell their horticultural crops at very low prices for their customers. Furthermore, the lack of ability in business planning, lack of marketing knowledge and the perish ability of their products contribute to their weak influential position in the supply chain (Wei et al., 2001).

The use of chemicals must be regulated by appropriate health or food authorities. At the same time processing equipment used must be specific and standard for the treatment (McLauchlan and Bagshaw, 2001). However, in the present study rather than chemical, principally traditional techniques were more used. The treatment or the curing of root and tuber crops such as sweet potatoes and potatoes is an important practice if these crops are to be stored for long time (Kitinoja and Kader, 2002). Some plant materials are also useful as natural pesticides. For example the cassava leaves protect the harvested cassava roots from pests when used as packing material in boxes or bags during transport and short-term storage. The ashes of the leaves of *Lantana* spp. and *Ochroma logopur* have been used as a dust against aphids attacking stored potatoes. Neem leaves are also becoming more widely known and used throughout the world due to powerful pesticidal effect on food crops and non-toxicity to humans (Kitinoja and Kader, 2002).

Throughout the period between harvest and consumption, temperature control has been found to be the most important factor in maintaining product quality (Kitinoja and Kader, 2002). Keeping products at their lowest safe temperature will increase storage life by lowering respiration rate, decreasing sensitivity to ethylene gas and reducing water loss.

Exposure to the sun should be avoided as much as possible after harvest. After harvest cooling of horticultural crops is used to remove field heat. The delay in cooling will shorten post-harvest life and reduce quality. Field container should also be placed in the shade or loosely covered with leafy plant materials, straw or an inverted empty container (Kitinoja and Kader, 2002). However, repeated cooling and warming of horticultural products cause deterioration. Keeping products too cool can cause serious problem such as chilling (Shewfelt, 1986).

Although post-harvest loss estimate for horticultural crops was difficult, the current record was comparable with estimated losses recorded in Nigeria (Idah et al., 2007). Furthermore, pest, drought, shortage of fertilizer, and price of fuel for pumping water as the major constraints of horticulture production in Eastern Ethiopia were identified byBezabih and Hadera (2007). The estimated post-harvest losses of fruits and vegetables range from 20% to 40% (Wiersinga and de Jager, 2009). Increased returns to growers might come from proper management after harvest (Wiersinga and de Jager, 2009).

The increasing horticulture production can contribute to the commercialization of the rural economy and creates many jobs. However, expanding the scale of horticulture production is often hindered by lack of market access, market information and many biological factors (Abay, 2007). Bezabih and Hadera, (2007) also argued seasonal production to be inversely related to price.

Information on price, product demand, product supply, market place and buyers and sellers should be gathered. Although horticultural crops are exported to Djibouti and Somalia, they are not as such standardized. The logistics such as handling, packing, loading and unloading, storage, road and transportation are poor. This results in selling the Ethiopian horticultural products to conventional traditional prices (Wiersinga and de Jager, 2009).

In general, the mishandling during harvesting, packaging, transportation and storage and unfavorable climatic condition and contamination are causing mechanical, pathological and physiological damage. The support given for the improvement and reduction of post-harvest loss and quality deterioration of horticultural crops from concerned bodies is low and insufficient. Therefore urgent intervention to the existing problems is highly required. In addition, an effective and efficient intervention policies and strategies need to be developed.

ACKNOWLEDGMENTS

Our appreciation goes to Dr. Fekedu Lemessa, Paulos Asrat, and Getinet Tolessa for their continuous follow-up, constructive comments, encouragements and facilitation of this study. We express our gratitude to all concerned Dire Dawa Administration officers in particular to Workineh Tilahun and Abdi Mohammed, Tesfaye Alemayehu and Ketema Zeleke to enumerators and respondents. We greatly recognize the financial, material and equipment support by Dire Dawa University and Dire Dawa Rural and Agriculture Developmental Office.

REFERENCES

1. Abay, A., 2007. Vegetable market chain analysis in Amhara National Regional State: the case of Fogera Woreda, South Gondar Zone. M.Sc. Thesis. Haramaya University.

2. Abdel-Rahman, A.M., 2012. Effect of cement dust pollution on the yield and quality of Ficus carica L. Fruits J. Life Sci. 6, 276–282.

3. Abdel-Rahman, A.M., Ibrahim, M.M., 2012. Effect of cement particulates deposition on eco-physiological behaviors of Halophytes in the salt marshes of Red Sea. Am.-Euras. J. Agric. Environ. Sci. 12, 490–498.

4. Alazar, A., 2007. Horticultural Marketing in Ethiopia. M.Sc. Thesis. Haramaya University, Haramaya.

5. Asubiojo, O.I., Aina, P.O., Oluwole, A.F., 1991. Effect of cement production on the elemental composition of soil in the neighbouhood of tow cement factories. Water Air Soil Pollut. 57 (58), 819– 828

6. Awole, S., Woldetsadik, K., Workneh, T.S., 2011. Yield and storability of green fruits from hot pepper cultivars (Capsicum spp.). Afr. J. Biotechnol. 10, 12662–12670.

7. Bekele, W., 1989. Horticulture marketing systems in Ethiopia. First international symposium on horticultural economics in developing countries. Acta Hortic. 1, 21–31.

8. Bezabih, E., Hadera, G., 2007. Constraints and problems of horticulture production and marketing in Eastern Ethiopia. Dryland Coordination Report, G46, Osolo, p. 91.

9. Bureau of Finance and Economic Development (BFED), 2010. Dire Dawa Administration Statistical Abstract 2005/06–2008/09. BFED, Policy Study and Plan Development Co-operation Office, Dire Dawa.

10. Central Statistical Agency (CSA), 2008. Summery and Statistical Report of the 2007 Population and Housing Census: Population size by Age and Sex. Federal Democratic Republic of Ethiopia, Population Census Commission, Addis Ababa, Ethiopia.

11. Choudhury, M.L., 2006. Recent developments in reducing postharvest losses in the Asia–Pacific Region. In: Rolle, R.S. (ed.), Postharvest Management of Fruit and Vegetables in the AsiaPacific Region. Asian Productivity Organization, and Food and Agriculture Organization of the United Nations, Tokyo, p. 312.

12. Debele, D., Fekadu, L., Ravishankar, H., 2007. Development and performance evaluation of indigenously made cooling chambers for extending the shelf life of mangoes and sweet oranges. East Afr. J. Sci. 1, 127–135.

13. Fakhry, A.M., Migahid, M.M., 2011. Effect of cement-kiln dust pollution on the vegetation in the western Mediterranean desert of Egypt. World Acad. Sci. Eng. Technol. 57, 28–34.

14. Fantahun, A., Williamson, S., 2001. Ethiopian farmers test alternatives to pesticides. Pestic. News 52, 8–9.

15. Food and Agriculture Organization (FAO), 1989. Prevention of PostHarvest Food Losses: Fruit, Vegetables and Root Crops; A Training Manual. UNFAO, Rome.

16. Idah, P.A., Ajisegiri, E.S.A., Yisa, M.O., 2007. Fruits and vegetables handling and transportation in Nigeria. Aust. J. Technol. 10, 175– 183.

17. Kitinoja, L., Kader, A.A., 2002. Small-Scale Postharvest Handling Practices: A Manual for Horticultural Crops, fourth ed. University of California, Postharvest Technology Research and Information Center, Davis.

18. Liu, J.L., Du, M., Shang, K.Y., Chen, S.Y., Xu, H.B., Xie, M.Y., 1997. The effects of cement dust pollution on rice, rape and soil. J. Plant Resour. Environ. 6, 42–47.

19. McLauchlan, R., Bagshaw, J., 2001. Australian postharvest technologies for fresh fruits and vegetables. In: O'Hare, T., Bagshaw, J., Li, W., Johnson, G.I., (Eds.), Postharvest Handling of Fresh Vegetables. Proceedings of a Workshop Held in Beijing, People's Republic of China, 9–11 May 2001. ACIAR Proceedings No. 105.

20. Milaku, J., 2005. Patterns and determinants of fruit and vegetable demand in developing countries: a multi-country comparison (Ethiopia). Fruit and Vegetables for Health. Report of a Joint FAO/WHO

Workshop, 1–3 September 2004, Kobe. Moy, G., 2005. Food safety aspects in fruit and vegetables. Fruit and Vegetables for Health (FAO/WHO). Report of a Joint FAO/WHO Workshop, 1–3 September 2004, Kobe.

21. Mrema, C.G., Rolle, S.R., 2002. Status of the postharvest sector and its contribution to agricultural development and economic growth. In: 9th JIRCAS International Symposium – Value Addition to Agricultural Product, pp. 13–20.

22. Olayemi, F.F., Adegbola, J.A., Bamishaiye, E.I., Daura, A.M., 2010. Assessment of post-harvest challenges of small scale farm holders of tomatoes, bell and hot pepper in some local government areas of Kano State, Nigeria. Bayero J. Pure Appl. Sci. 3, 39–42.

23. Raajasubramanian, D., Sundaramoorthy, P., Baskaran, L., Sankar, K., Ganesh, A.L., Chidambaram, A., Jeganathan, M., 2011. Effect of cement dust pollution on germination and growth of groundnut (Arachis hypogaea L.). Int. Multidisc. Res. J. 1, 25–30.

24. Saralabai, V.C., Vivekanandan, M., 1992. Positive effect of cement kiln-exhaust on Legume crop under simulation study. Appl. Biochem. Biotechnol. 36, 36–46.

25. Saxena, K.N., Pala-Okeyo, A., Seshu-Reddy, K.V., Omolo, E.O., Ngode, L., 1990. Insect pest management and socio-economic circumstances of small-scale farmers for food crop production in western Kenya: a case study. Insect Sci. Appl. 10, 443–462.

26. Shewfelt, R.L., 1986. Postharvest treatment for extending shelf life of fruits and vegetables. Food Technol. 40, 7078–7089.

27. Singh, A., Singh, Y., 1992. Effect of vibration during transportation on the quality of tomatoes. J. Agric. Mech. Asia Afr. Latin Am. 23, 70–72.

28. Singh, S.N., Rao, D.N., 1980. Growth of wheat plants exposed to cement dust pollution. Water, Air Soil Pollut. 14, 241–249.

29. Sirivatanapa, S., 2006. Packaging and transportation of fruits and vegetables for better marketing. In: Rolle, R.S. (Ed.), Postharvest Management of Fruit and Vegetables in the Asia-Pacific Region. Asian Productivity Organization, and Food and Agriculture Organization of the United Nations, Tokyo, p. 312.

30. Thompson, J.F., Mitchell, F.G., Runsey, T.R., Kasmire, R.F., Crisosto, C.H., 1998. In: Commercial Cooling of Fruits, Vegetables and Flowers, vol. 1211. DANR Publications, pp. 61–68.

31. Wei, S., Woods, E., Adar, D., Singgih, S., 2001. West Timor mandarin marketing case study implications for supply-chain management in developing countries. In: O'Hare, T., Bagshaw, J., Li, W., Johnson, G.I. (Eds.), Postharvest Handling of Fresh Vegetables. Proceedings of a Workshop Held in Beijing, People's Republic of China, 9–11 May 2001. ACIAR Proceedings No. 105, p. 126.

32. Wiersinga, R., de Jager, A., 2009. In: Business Opportunities in the Ethiopian Fruit and Vegetable Sector. Wageningen University and Research Centre – LEI, p. 46.

33. Yohannes, A., 1989. Economics of horticultural production in Ethiopia: first international symposium on horticultural economics in developing countries. Acta Hortic. 16, 15–19.

Chapter 3

Sound waves delay tomato fruit ripening by negatively regulating ethylene biosynthesis and signaling genes

Joo-Yeol Kim[a, 1], Jin-Su Lee[b, 1], Taek-Ryoun Kwon[a], Soo-In Lee[a], Jin-A. Kim[a], Gyu-Myoung Lee[a], Soo-Chul Park[a], Mi-Jeong Jeong[a, ,]

a Department of Agricultural Biotechnology, National Academy of Agricultural Science (NAAS), 370 Nongsaengmyoeng-ro, Wansan-gu, Jeonju, Jeollabuk-do 560-500, South Korea

b Postharvest Research Team, National Institute of Horticultural and Herbal Science (NIHHS), 100, Nongsaengmyeong-ro, Iseo-myeon, Wanju-gun, Jeollabuk-do 565-852, South Korea

HIGHLIGHTS

- Tomato treated with sound waves of 1 kHz delayed ripening compared to control.
- Ethylene production and respiration rate were lower in the sound treated tomatoes.
- Changes in surface color from green to red were delayed in the treated fruits.
- mRNA level of ethylene biosynthetic genes was very low in the sound treated fruits.

ABSTRACT

Regulation of tomato fruit ripening may help extend fruit shelf life and prevent losses due to spoilage. Here, tomato fruit were investigated whether sound treatment could delay their ripening. Harvested fruit were treated with low-frequency sound waves (1 kHz) for 6 h, and then monitored various characteristics of the fruit over 14-days at 23 ± 1 °C. Seven days after the treatment, 85% of the treated fruit were green, versus fewer than 50% of the non-treated fruit. Most of the tomato fruit had transitioned to the red ripening stage by 14 days after treatment. Ethylene production and respiration rate were lower in the sound-treated than non-treated tomatoes. Furthermore, changes in surface color and flesh firmness were delayed in the treated fruit. To investigate how sound wave treatment effects on fruit ripening, the expression of ethylene-related genes was analyzed by quantitative real-time RT-PCR analysis. The expression level of several ethylene biosynthetic (*ACS2, ACS4, ACO1, E4* and *E8*) and ripening-regulated (*RIN, TAGL1, HB-1, NOR, CNR*) genes was influenced by sound wave treatment. These results indicated that sound wave treatment delays tomato fruit ripening by altering the expression of important genes in the ethylene biosynthesis and ethylene signaling pathways.

KEYWORDS

Sound waves; Tomato; Ethylene biosynthesis; Fruit ripening; Ethylene signaling

1. INTRODUCTION

Plants are frequently exposed to abiotic and biotic environmental stresses. Stress signals are transmitted to cells, where they activate various stress response mechanisms. The mechanism by which plants recognize external signals and then alter the expression of genes to affect physiological and/or metabolic processes is complex. Much research has examined the signal transduction mechanisms that regulate the defense or growth responses of plants following artificial stimulation, such as mechanical stimulation (Lin et al., 2009, Braam, 2004 and Monshausen and Gilroy, 2009). Another such external signal is sound.

Previous reports on the effects of sound on plants failed to specify the exact experimental conditions (Weinberger and Measures, 1978). Sound vibrations stimulate seed germination and plant growth and increase the rate of hypocotyl elongation in *Oryza sativa* (rice) and *Cucumis sativus* (cucumber) seedlings (Takahashi et al., 1992) and in Arabidopsis (Johnson et al., 1998). Jomdecha and Prateepasen (2011) showed that low-intensity ultrasonic sound enhances the growth of *Saccharomyces cerevisiae.*Furthermore, transgenic rice plants harboring an *Alcohol dehydrogenase* (*Ald*) promoter:*GUS* reporter construct showed a significant increase in *GUS* expression after sound treatment at 125 Hz, 250 Hz, or 1 kHz. By contrast, these plants showed a significant decrease in *GUS* mRNA levels after a treatment of 50 Hz, suggesting that the*Ald* promoter responds to sound in a frequency-specific manner (Jeong et al., 2008). In addition, through biochemical analyses, Bochu et al. (2004) reported that sound wave stimuli (1.4 kHz, 95 dB, 10 days) were associated with increased levels of indole acetic acid (IAA) and decreased levels of abscisic acid (ABA) in Chrysanthemum. Accordingly, sound waves are thought to regulate a variety of plant hormones and genes.

Fruits are divided into climacteric and non-climacteric types based on their respiration patterns during ripening. Climacteric fruits, such as tomato, apple, avocado, melon, pear, kiwifruit, and peach, exhibit increased respiration and ethylene production during ripening. However, respiration and ethylene production rates barely change during ripening of non-climacteric fruit, such as grapes, cherry, strawberry, pineapple, and citrus (Lin et al., 2009 and Cara and Giovannoni, 2008). Almost all fruits start aging rapidly when ethylene production increases and once fruit have ripened their quality declines. Suppression of senescence in fruit results in a longer shelf life. Tomato is a major model system for the study of the molecular mechanisms underlying fruit development and the role of ethylene in climacteric fruit ripening. It has various useful characteristics, such as efficient genetic transformation, distinct ripening phenotypes, small genome size, short generation time, and simple diploid genetics (Liu et al., 2014 and Giovannoni, 2004).

Ethylene is a plant hormone involved in fruit, flower, and leaf senescence. The ethylene biosynthesis precursor S-adenosylmethionine (SAM) is converted into 1-aminocyclopropane-1-carboxylic acid (ACC) by ACC synthase (ACS), and ACC oxidase (ACO) then catalyzes the synthesis of ethylene (Yang and Hoffman, 1984 and Johnson and Ecker, 1998). Ethylene biosynthesis in tomato fruit can be divided into systems 1 and 2 during climacteric fruit ripening (Lelièvre et al., 1997 and Alexander and Grierson,

2002). System 1 is auto-inhibitory and functions during normal growth and development and the stress response, whereas system 2 operates during floral and leaf senescence and fruit ripening (Katz et al., 2004 and Barry and Giovannoni, 2007).

Expression of *LeACS2, LeACS4,* and *LeACO1* increases significantly during fruit ripening, indicating that these genes are major regulators of ripening (Hoogstrate et al., 2014 and Lincoln et al., 1993). The ethylene-inducible genes *E4* and *E8* also have important functions during fruit ripening. Transcription of *E4* is rapidly activated by ethylene in both leaves and fruit. By contrast, expression of *E8* is highly induced by ethylene in tomato fruit, but is not induced in response to ethylene in leaves. Although *E4*and *E8* are coordinately expressed during fruit ripening, *E8* is controlled both by ethylene and by separate developmental signals, whereas *E4* expression is strictly dependent on ethylene (Kesanakurti et al., 2012 and Lincoln and Fischer, 1988). Furthermore,*TOMATO AGAMOUS-LIKE 1* (*TAGL1*) MADS box gene as a transcription factor plays an important function in both fruit expansion and ripening. Silencing of *TAGL1* does not affect floral organs, but affects ripening of the tomato fruit (Vrebalov et al., 2009,Giménez et al., 2010 and Pan et al., 2010). *RIPENING-INHIBITOR (RIN)* functions upstream of ethylene in the ripening cascade and determine the competence of fruit to ripen. *RIN* was shown to regulate *LeACS2* and *LeACS4* expression (Vrebalov et al., 2002). Consequently, *TAGL1* and *RIN* affect ethylene metabolism (Yokotani et al., 2004). In addition, virus-induced gene silencing of *LeHB-1,* which encodes an HD-Zip homeobox protein, reduced *LeACO1* mRNA expression and inhibited ripening in tomato fruit (Lin et al., 2008). In this study, tomato fruits were investigated whether sound wave treatment could delay ripening by regulating the expression of genes involved in ethylene biosynthesis and signaling.

2. MATERIALS AND METHODS

2.1. Plant materials and treatment conditions

Tomato (*Solanum lycopersicum* L. cv. Dotaerang) fruits were sampled at the mature green stage of development (average fruit weight 172.9 ± 1.7 g) from a commercial greenhouse facility at Yong-in, Korea. After transport to the laboratory of the Rural Development Administration on the day of harvest, the fruits were exposed to a 1 kHz sound wave for 6 h. The single-frequency signal was generated using Pro Tools M-Powered software (Avid

Technology, USA). The speaker volume was 100 dB. To prevent extraneous noise, experiments were conducted inside a custom-made, sound-proof chamber (Korea Scientific Technique Industry Co., Korea). The sound level within the chamber was approximately 40 dB, while the sound level in a commercial growth chamber reaches generally about 80 dB. Tomato fruits were placed in the sound-proof chambers to prevent transfer of vibrations between samples during the sound wave treatments. The sound wave treatment was conducted at 23 ± 1 °C. Control groups were achieved with the exposure of fruit in the same conditions except for the sound interference. Fruit samples were placed in a storage room at 23 ± 1 °C and monitored for quality changes after sound wave treatments.

2.2. Ethylene production and respiration rate determination

Thirty-two tomato fruits per treatment were used for assessing ethylene production and respiration rate at each measurement interval. Eight tomato fruits were weighed and sealed in a 3900 mL plastic container (n = 4) equipped with a septum for 1 h. Headspace samples (1 mL) were withdrawn from the container and ethylene and carbon dioxide concentrations were analyzed using a gas chromatograph (Varian 450, Varian Inc., CA, USA) equipped with a flame ionization detector (FID) and thermal conductivity detector (TCD).

Ethylene levels were measured by a FID equipped with an active alumina 60–80 mesh column. The injector, oven, and detector were maintained at 110, 70, and 250 °C, respectively. Respiration rate was determined based on carbon dioxide levels using a TCD equipped with an active carbon 60–80 mesh column. The injector, oven, and detector were maintained at 110, 70, and 150 °C, respectively. Helium (30 mL min^{-1}) was used as the carrier gas for ethylene and carbon dioxide detection.

2.3. Fruit quality parameters

Surface color was measured using a Chromameter (CR-400, Konica Minolta Sensing Inc., Osaka, Japan). Two equatorial regions of each fruit were marked and the surface color of 30 fruit per treatment (n = 30) was assessed repeatedly at each measurement interval. The hunter '*a*' value ranging from green to red was used to determine surface color change during storage.

Flesh firmness was determined using a texture analyzer (TA plus, Ametek Lloyd Instruments Ltd., Hampshire, UK) equipped with a 5 mm cylindrical probe and 250 N load cell. The equatorial region of each fruit was cut into 20-mm thick slices and flesh firmness was evaluated. The probe was pressed 10 mm into the mesocarp surface of a slice with a crosshead speed of 2 mm s^{-1} and the positive peak force (*N*) was recorded. Flesh firmness of slices derived from 20 fruits per treatment ($n = 20$) were evaluated at each measurement interval.

2.4. RNA extraction and quantitative real-time PCR (qRT-PCR)

Tomato fruits were obtained from at least three individuals per step. Directly after harvesting, the fruits were frozen in liquid nitrogen and stored at −80 °C. These fruits were ground into powder with pestle and mortar in liquid nitrogen to keep the tissue frozen. Total RNA was extracted using a Plant RNeasy Extraction Kit (Qiagen). Total RNA was treated with DNase I (Qiagen) and cDNA was generated using ampfiRivert Platinum cDNA Synthesis Master Mix (GenDEPOT, South Korea). Quantitative real-time RT-PCR (qRT-PCR) was performed using SYBR Premix Ex Taq (Takara) and the StepOnePlus Real-Time PCR System (Applied Biosystems). The relative mRNA levels were determined by normalizing the PCR threshold cycle number of each gene with that of the*Ubiquitin3* reference gene. Three biological replicates were used in the experiments. The primers used for PCR are shown in Table S1.

3. RESULTS AND DISCUSSION

3.1. Sound wave treatment delayed ripening in tomato fruit

Sound wave significantly effects on the ripening process of tomato fruit. To evaluate the impact of sound waves on fruit ripening, tomato fruits at the mature green stage were treated with sound waves of 0 Hz, 250 Hz, 500 Hz,

800 Hz, 1 kHz, and 1.5 kHz for 6 h. For all of the sound wave treatments except for 800 Hz and 1.5 kHz, ripening was delayed as compared with the non-treated control (Fig. 1). Since the 1 kHz sound wave showed the strongest effect on ripening process, tomato fruit were treated with 1 kHz sound for phenotypic analysis over 14 days of storage at 23 ± 1 °C. Thus, most of the results in this study were obtained using tomato fruit treated with 1 kHz, unless otherwise indicated. Under these conditions, the mature green stage tomato fruit transitioned to the breaker stage after 2 days. Tomato fruit treated with sound waves of 1 kHz exhibited delayed ripening as compared with the non-treated fruit at 5 or 7 days after treatment (Fig. 2A). Seven days after treatment, over 85% of the treated fruit was still green, whereas less than 50% of the non-treated tomato fruit were green (Fig. 2B). Finally, most of the tomato fruit transitioned to the red ripening stage by 14 days after treatment. These results indicate that sound wave treatment delays tomato fruit ripening.

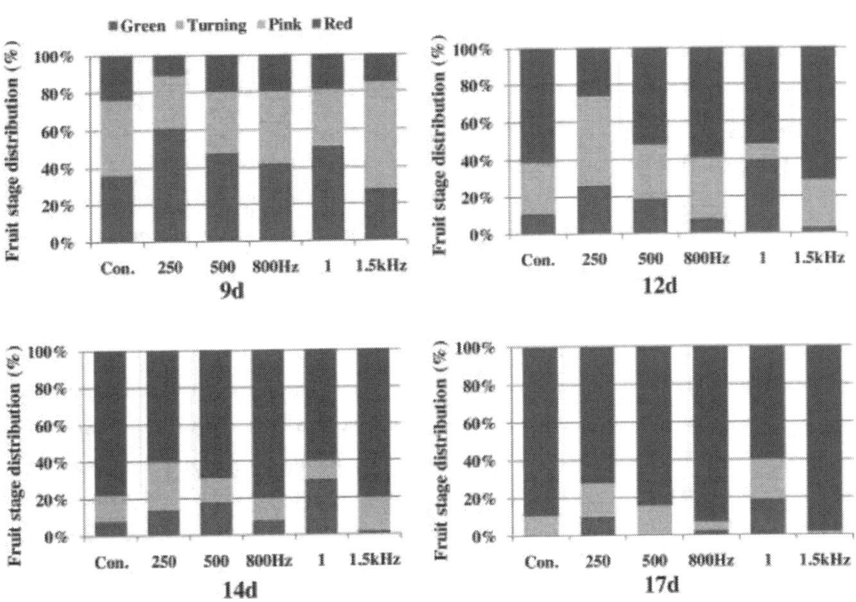

Figure 1. Comparison of fruit surface color change in response to various sound treatments in tomato. Phenotypic analysis of tomato fruit ripening at 9, 12, 14, and 17 days after a 6-h sound treatment (non-treatment, 250 Hz, 500 Hz, 800 Hz, 1 kHz, and 1.5 kHz). Fruit were maintained at 23 °C throughout the experiment. The average results of three experiments are shown.

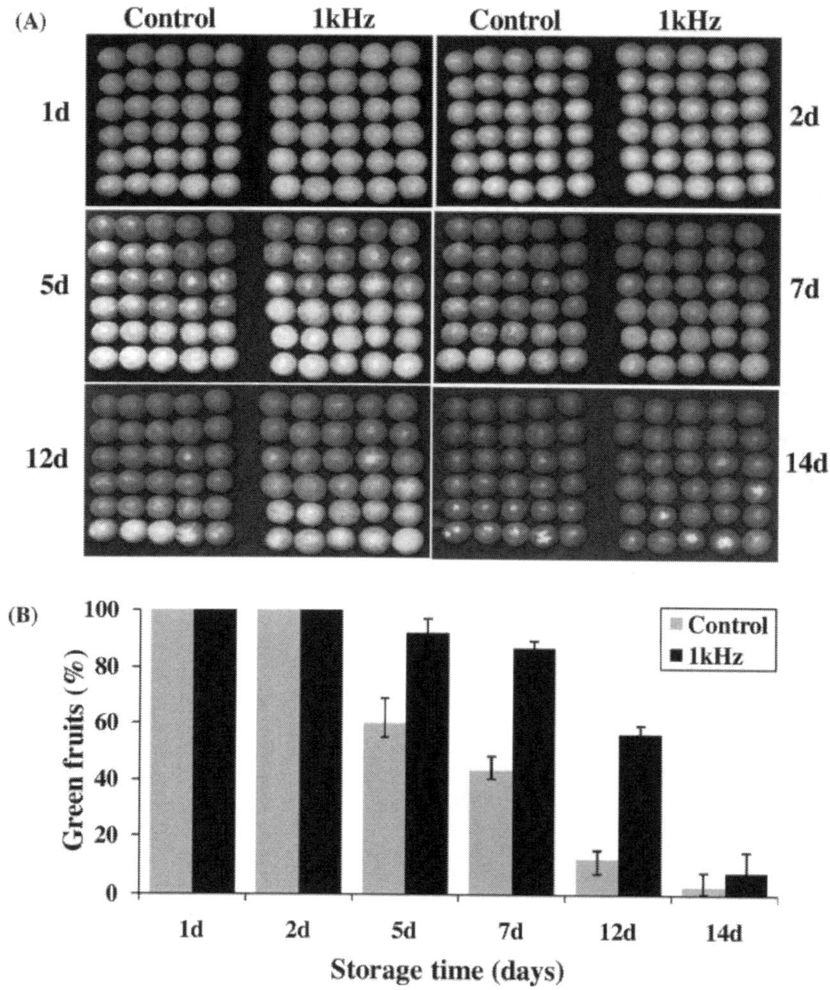

Figure 2. Ripening delay in tomato fruit subjected to a sound treatment of 1 kHz. (A) Phenotypic analysis of tomato fruit ripening at 1, 2, 5, 7, 12, and 14 days (at 23 ± 1 °C) after sound treatment (1 kHz), (B) the percentage of green tomato fruit during the 14 days (at 23 ± 1 °C) following sound treatment (1 kHz). Error bars represent SD of three biological replicates.

These findings could be explained partially with the previous reports on the effects of sound wave to the deformability of plant cell. Sound wave exposure changes the deformability of the plant cell membrane, and the deformability is closely related to the frequency of the sound wave (Bochu et al., 2001).

3.2. Sound wave treatment reduced ethylene production and respiration rate in tomato fruit

External stimuli like sound wave also could regulate various biological process and genes. A scientific report by Johnson et al. (1998) reveals that mechanotransduction occurs in cells and tissues, and external mechanical stresses stimulate various signal transduction mechanisms. In this study, external stimuli with the sound wave of 1 kHz applied to 32 tomato fruits at various time points after treatment (1–15 days) prior to the determination of ethylene production. As shown in Fig. 3A, ethylene production was lower in tomatoes treated with sound waves of 1 kHz than in the non-treated control. The ethylene production was observed the greatest reduction at 12 days after treatment. These results indicate that the imposition of the sound treatment wave significantly effects on the biological process of ethylene, resulting in the delay of the fruit ripening. These findings agree well with the previous postharvest physiology of tomato fruit suggested byLin et al., (2009) and Cara and Giovannoni (2008). The ripening of tomato fruit, as a climacteric type fruit, is associated with increased respiration and ethylene production.

Next, the respiration rate was measured in tomato fruit treated or not with sound waves. To examine the respiration rate, the evolution rate of carbon dioxide was analyzed using a gas chromatography. The respiration rate was lower in the treated tomato fruit one day after treatment than in the non-treated control, and the greatest relative reduction was observed at 12 days after treatment (Fig. 3B). In the non-treated fruit, the respiration rate peaked at 12 days, and then decreased rapidly. At this stage, the ripening stage seemed to be completed, and senescence seemed to be in progress. By contrast, the treated tomato fruit appeared still to be ripening, because the respiration rate increased even at 15 days after treatment. These results indicate that sound wave treatment (1 kHz) delays ripening of tomato fruit by controlling the ethylene production as well as respiration rate.

3.3. Sound wave caused changes in surface color and hardness in tomato fruit

Surface color is the most important external characteristic for visual assessing of ripeness in tomato fruit. The degree of ripening is usually estimated using color charts. Surface color analysis using a chromameter provides an objective way to quantitatively measure and compare the color of tomato fruit by standardizing the elements necessary to view color. The hunter '*a*' value, which denotes colors ranging from green to red, was used to determine surface color changes during storage.

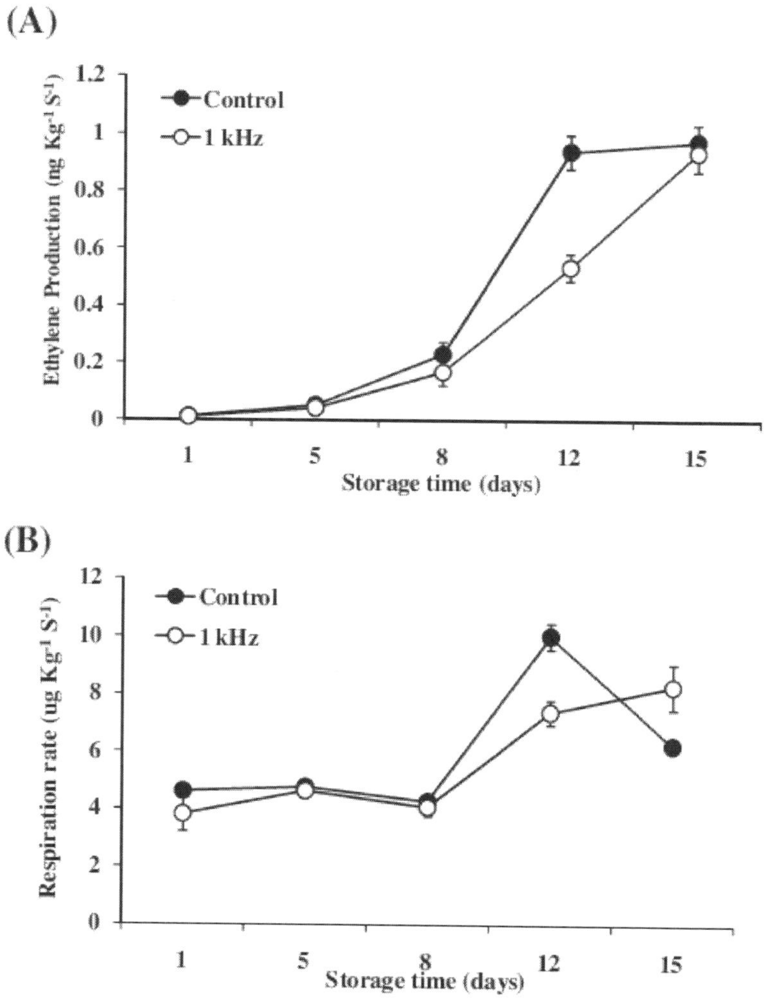

Figure 3. The effect of sound treatment on ethylene production and respiration rate in tomato fruit. (A) Changes in ethylene production of tomato fruit during storage at 23 °C after sound treatment (1 kHz). (B) Changes in respiration rate of tomato fruit during storage at 23 °C after sound wave treatment (1 kHz). Error bars represent SD of three biological replicates.

Tomato fruits were treated with 1 kHz sound waves for 6 h and then measured the surface color at various time points during the next 16 days (at 23 ± 1 °C). Changes in surface color from green to red were delayed in the treated tomato fruit compared with the non-treated fruit (Fig. 4A).

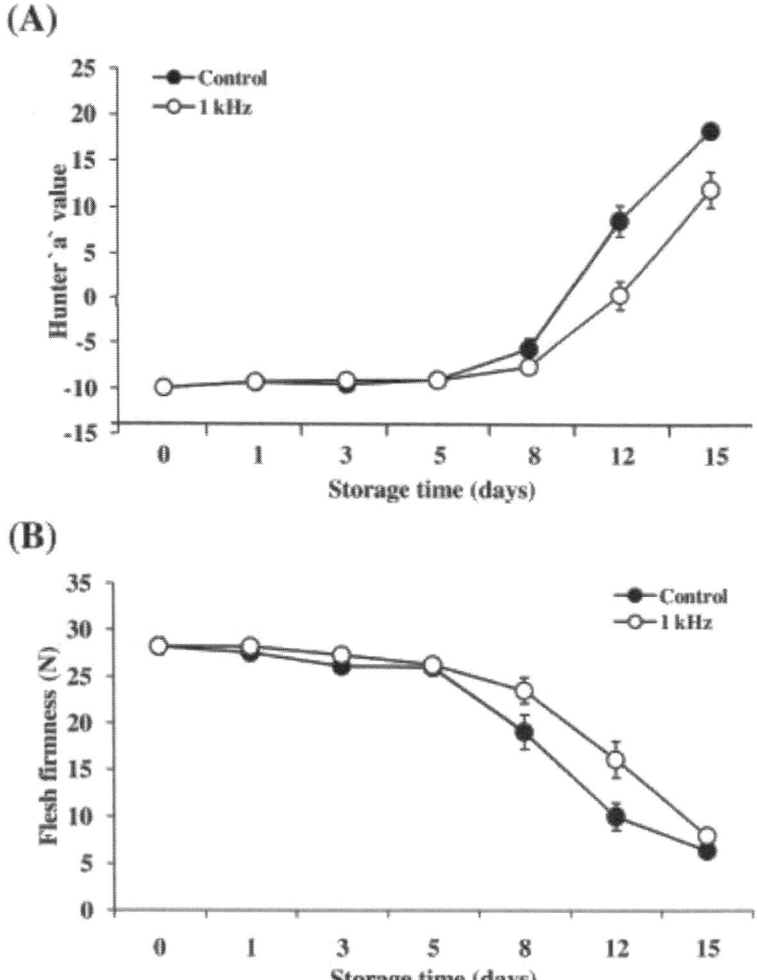

Figure 4. Sound treatment causes changes in surface color and hardness in tomato fruit. (A) Changes in Hunter '*a*' values of tomato fruit surfaces during storage at 23 °C after sound signal treatment. a* (red–green) axis: positive values are red; negative values are green; 0 is neutral. (B) Changes in flesh firmness of tomato fruit during storage at 23 °C after sound signal treatment. Samples were measured in biological triplicate. Error bars represent SD.

To further determine the effect of sound wave treatment on tomato fruit ripening, examined whether sound wave treatment influenced flesh firmness. The flesh firmness of non-treated tomato fruit decreased sharply after 5 days-long storage in this study, whereas that of fruit treated with sound waves decreased at a more gradual rate (Fig. 4B). These results demonstrate that sound wave treatment helped to maintain their flesh firmness.

Our study showed that the ethylene production and respiration rate were significantly reduced in plants exposed to a 1 kHz sound wave treatment (Fig. 3A and B). In addition, changes in surface color from green to red were delayed in sound wave-treated tomato fruit (Fig. 4A) and the flesh was firmer in the treated tomato fruit than in the non-treated fruit (Fig. 4B).

Thus, sound wave treatment delayed the ripening of tomato fruit (Fig. 1 and Fig. 2). Because the color change was delayed and ethylene production and rate of respiration were reduced in the sound wave-treated tomato fruit, the expression of genes involved in ethylene biosynthesis was analyzed.

3.4. Sound wave treatment affects ethylene biosynthesis-related genes

This study found that sound wave controls the expression of genes involved in the biochemical processes associated with ripening. Ethylene mediates certain developmental processes, most notably early fruit development and ripening. Having established that sound wave treatment delayed tomato fruit ripening (Fig. 1, Fig. 2, Fig. 3 and Fig. 4), tomato fruit ripening assumed to be influenced ethylene-related responses by sound wave treatment. Therefore, the expression pattern of major genes involved in ethylene biosynthesis was analyzed after sound wave treatment. The transcript levels of ethylene biosynthetic genes in sound wave-treated tomato fruit was examined at various time points after treatment (at 23 ± 1 °C) via quantitative real-time RT-PCR analysis with the gene-specific primers listed in Table S1. Under normal conditions, most of the ethylene biosynthetic genes and ripening regulated genes in tomato were expressed at low levels in green fruit, but the transcription levels of each of these genes increased at the onset of ripening, as the fruit started to produce ethylene (Fig. 5). As shown in Fig. 5, in compare to control, the expressions of *LeACS2, LeACS4* and *LeACO1* (ethylene biosynthesis-related genes) were reduced in sound wave treated tomatoes. The difference of expression level of these genes between treated and untreated groups substantially increased from 7 to 14 days. The expression level of *E4* and *E8* genes (ethylene-inducible genes) were also significantly reduced by the sound wave treatment. These genes associated with ethylene biosynthesis exhibited the most striking changes in expression between 5 and 7 days after the sound treatment. In particular, the expression of *LeACS2, LeACS4,* and *LeACO1* was markedly reduced by sound treatment, and this reduction in expression is expected to reduce ethylene production. The expression level of *E4* and *E8* genes (ethylene-inducible genes) were also significantly reduced by the sound wave treatment (Fig. 5), probably due to the low concentrations of ethylene in sound-treated tomato fruit. These results well agree with the previous report on the strict dependence of those target genes with ethylene during fruit ripening (Kesanakurti et al., 2012 and Lincoln and Fischer, 1988).

3.5. Sound wave treatment triggers expression of ripening-related genes

RIN, LeHB-1, CNR, and *NOR* encode transcription factors that regulate the expression of several genes responsible for ethylene biosynthesis. *RIN* belongs to the MADS-box family of transcription factors, which are important regulators of vegetative growth, flowering time, and floral development (Vrebalov et al., 2002). *LeHB-1*, a class-I HD-Zip gene, encodes a transcription factor that binds to the promoters of ethylene biosynthesis-related genes that have recently been identified (Lin et al., 2008). In addition, the*TOMATO AGAMOUS-LIKE 1* (*TAGL1*) MADS box gene encodes a transcription factor that plays an important function in both fruit expansion and ripening. *TAGL1* promotes ripening by activating *ACS2*, and is reported to be a target of *RIN* (Itkin et al., 2009 and Vrebalov et al., 2009). Silencing of *TAGL1* or *RIN* delays ripening (Vrebalov et al., 2009, Giménez et al., 2010 and Pan et al., 2010). Recessive *rin* mutant tomatoes fail to ripen, as these mutants do not produce high levels of endogenous ethylene and do not respond to exogenous ethylene. *TAGL1* is expressed at early and later stages at the onset of ripening. Our results showed that the mRNA level of *RIN* decreased significantly as compared with the non-treated control fruit (Fig. 6). In fruit treated with sound waves, the expression level of *LeHB-1* was reduced after 1 day and 2 days, but increased after 5 days. Furthermore, *TAGL1* expression was reduced in the treated fruit, except between 7 and 12 days after sound wave treatment.

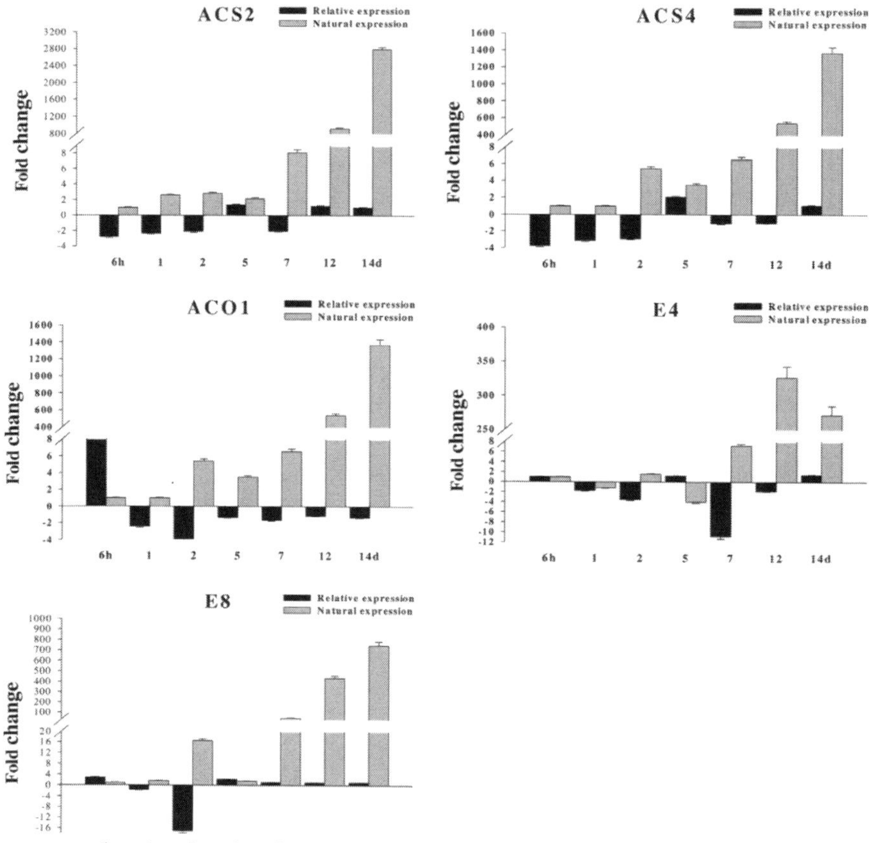

Figure 5. Expression of ethylene biosynthetic genes in tomato fruit after sound treatment. Expression of ethylene biosynthesis-related genes in tomato fruit at the indicated times (at 23 °C) after sound treatment (1 kHz) for 6 h, as determined by qRT-PCR. Natural expression is genes expression in non-treatment tomato fruit. Relative expression is value compared with natural expression in 1 kHz sound treatment tomato fruit. Error bars represent SD of three biological replicates.

Also, *CNR* and *NOR* have been reported to regulate the expression of genes related to fruit ripening, including ethylene and carotenoid biosynthesis (Giovannoni, 2004). In this study, the transcript level of *CNR* decreased at 1–3 days after sound wave treatment, but increased at 7 days after treatment. The expression pattern of *CNR* was opposite in the sound-treated and non-treated fruit (Fig. 6). Meanwhile, the expression of *NOR* was very low in sound-treated fruit compared to the non-treated fruit throughout the 14-day observation period (Fig. 6). These results showed that sound wave treatment delayed tomato fruit ripening by inhibiting the expression of some of the most important ethylene biosynthetic and ripening regulatory genes.

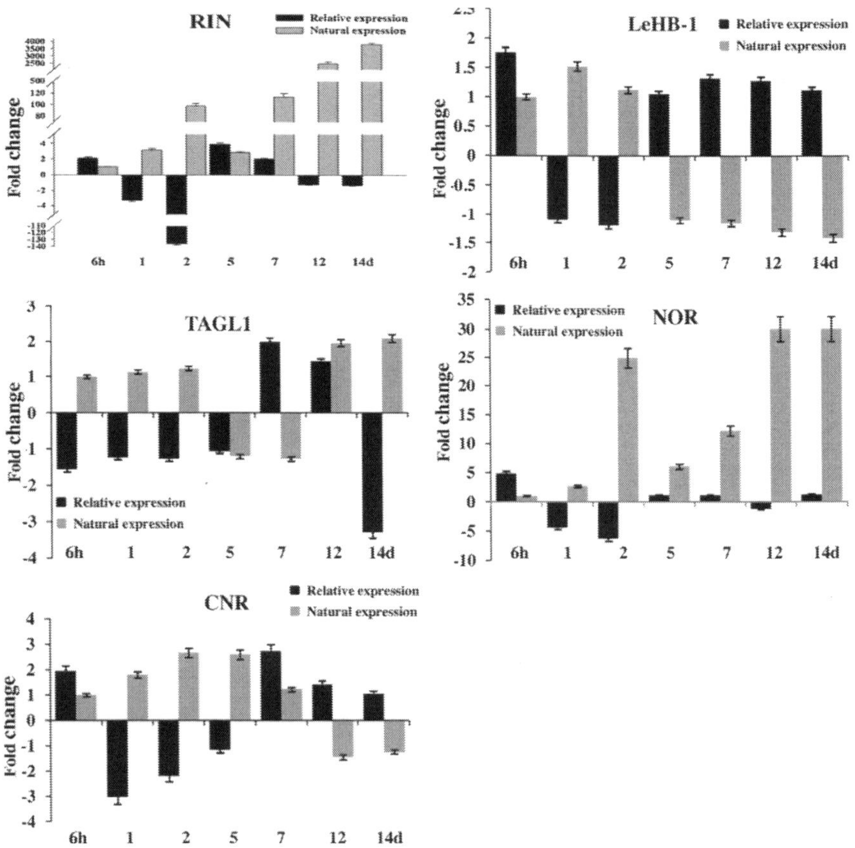

Figure 6. Expression of ripening-regulated genes in tomato fruit after sound treatment. Expression of ethylene response-related genes in tomato fruit at the indicated time points (at 23 °C) after sound treatment (1 kHz) for 6 h, as determined by qRT-PCR. Natural expression is genes expression in non-treatment tomato fruit. Relative expression is value compared with natural expression in 1 kHz sound treatment tomato fruit. Error bars represent SD of three biological replicates.

This study also showed that the mRNA levels of *RIN*, *ACS2,* and *ACS4* significantly increased after 7 days in non-treated fruit. However, the expression of these three genes was low, and remained low up to 14 days after sound wave treatment (Fig. 5 and Fig. 6).Martel et al. (2011) showed that *RIN* binds to the *ACS2* and *ACS4* promoter, but not to the*ACO1* promoter. *RIN* might be involved in reducing *ACS2* and *ACS4* expression. Due to the decreased expression of *RIN*, the expression of *ACS2* and *ACS4* may have been reduced, but the possibility cannot be ruled out that *ACS2* and *ACS4* expression is directly controlled by sound waves.

LeHB-1 has been reported to bind to the *LeACO1* promoter, which contains the putative HD protein binding sequence, and directly regulates a number of ripening-related genes (Lin et al., 2009). Manavella et al. (2006) showed that transgenic Arabidopsis plants harboring sunflower *Hahb-4* exhibited a remarkable delay in senescence and were less sensitive to ethylene. The expression of this gene suppresses the expression of genes associated with ethylene biosynthesis, such as *LeACO,* and with ethylene signaling, such as *ERF2* and *ERF5* (Manavella et al., 2006). Blume and Grierson (1997) reported that*LeHB-1* is highly expressed in mature green fruit and decreased at the breaker stage, whereas the expression of *LeACO1* increased in mature green fruit and accumulated during ripening. These results showed that the level of *LeHB-1* transcript declined in the green mature stage and increased after the breaker stage, whereas the expression level of *LeACO1* was higher in the early stage of ethylene biosynthesis and was maintained at low levels for up to 14 days after sound wave treatment, indicating that the expression pattern of these genes was affected by the sound wave treatment. As *LeHB-1* expression increased, the expression of *ACO1* was significantly reduced in sound wave-treated tomato fruit (Fig. 5 and Fig. 6).

TAGL1 is known to be involved in ripening in tomato (Itkin et al., 2009 and Vrebalov et al., 2009). *TAGL1* promotes ripening by activating *ACS2*, and is reported to be a target of*RIN* (Itkin et al., 2009 and Vrebalov et al., 2009). Silencing of *TAGL1* or *RIN* delays ripening (Vrebalov et al., 2009, Giménez et al., 2010 and Pan et al., 2010). Recessive*RIN* mutant tomatoes fail to ripen, as these mutants do not produce high levels of endogenous ethylene and do not respond to exogenous ethylene. *TAGL1* is expressed at early and later stages at the onset of ripening. These real-time RT-PCR results were similar to those reported by others (Busi et al., 2003 and Hileman et al., 2006) in non-treated fruit, but *TAGL1* expression was reduced during the early stage of ripening in tomato fruit treated with sound waves.

Every year, spoilage of postharvest agricultural products results in tremendous economic losses. The purpose of this study was to test the efficacy of delaying the ripening of tomato fruit by controlling the expression of genes related to ethylene biosynthesis through sound treatment. These results propose well that sound wave treatment delays tomato fruit ripening by regulating the expression of ethylene biosynthesis-related genes.

4. CONCLUSION

Findings in this study leads a conclusion that sound wave can be an external stimulus to delay the ripening of tomato fruit. This study investigated relevant molecular mechanisms through analysis of the expression of ethylene synthesis-related genes, controlling fruit ripening.

This study set the stage for further studies of sound wave-mediated responses in plants. Findings of this study indicate that acoustic biology can have viable applications in agriculture and may help to increase product quality and reduce losses. This study also provides the first indication that sound treatment can extend the ripening of tomato fruit. Since this area is still in the beginning stages, more detailed studies need to be performed in the future.

ACKNOWLEDGEMENT

This study was financially supported by NAAS Agenda Program (No. PJ0100252015 and PJ01049701) in Rural Development Administration, Jeonju,

REFERENCES

1. Alexander, L., Grierson, D., 2002. Ethylene biosynthesis and action in tomato: a model for climacteric fruit ripening. J. Exp. Bot. 53, 2039–2055. Barry, C.S., Giovannoni, J.J., 2007. Ethylene and fruit ripening. J. Plant Growth Regul. 26, 143–159.

2. Blume, B., Grierson, D., 1997. Expression of ACC oxidase promoter-GUS fusions in tomato and Nicotiana plumbaginifolia regulated by developmental and environmental stimuli. Plant J. 12 (4), 731–746.

3. Bochu, W., Hucheng, Z., Yiyao, L., Yi, J., Sakanishi, A., 2001. The effects of alternative stress on the cell membrane deformability of chrysanthemum callus cells. Colloids Surf. B 20, 321–325.

4. Bochu, W., Jiping, S., Biao, L., Jie, L., Chuanren, D., 2004. Sound wave stimulation triggers the content change of the endogenous hormone of the Chrysanthemum mature callus. Colloids Surf. B 37, 107–112.

5. Braam, J., 2004. In touch: plant response to mechanical stimuli. New Phytol. 165, 373–389.

6. Busi, M.V., Bustamante, C., D'Angelo, C., Hidalgo-Cuevas, M., Boggio, S.B., Valle, E.M., Zabaleta, E., 2003. MADS-box genes expressed during tomato seed and fruit development. Plant Mol. Biol. 52, 801–815.

7. Cara, B., Giovannoni, J.J., 2008. Molecular biology of ethylene during tomato fruit development and maturation. Plant Sci. 175, 106–113.

8. Giménez, E., Pineda, B., Capel, J., Antón, M.T., Atarés, A., Pérez-Martín, F., 2010. Functional analysis of the Arlequin mutant corroborates the essential role of the Arlequin/TAGL1 gene during reproductive development of tomato. PLoS One 5, e14427.

9. Giovannoni, J.J., 2004. Genetic regulation of fruit development and ripening. Plant Cell 16, S170–S180.

10. Hileman, L.C., Sundstrom, J.F., Litt, A., Chen, M., Shumba, T., Irish, V.F., 2006. Molecular and phylogenetic analyses of the MADS-box gene family in tomato. Mol. Biol. Evol. 23, 2245–2258.

11. Hoogstrate, S.W., van Bussel, L.J., Cristescu, S.M., Cator, E., Mariani, C., Vriezen, W.H., Rieu, I., 2014. Tomato ACS4 is necessary for timely start of and progression through the climacteric phase of fruit ripening. Front Plant Sci. 5 (466), 1–7.

12. Itkin, M., Seybold, H., Breitel, D., Rogachev, L., Meir, S., Aharoni, A., 2009. TOMATO AGAMOUS-LIKE 1 is a component of the fruit ripening regulatory network. Plant J. 60, 1081–1095.

13. Jeong, M.J., Shim, C.K., Lee, J.O., Kwon, H.B., Kim, Y.H., Lee, S.K., Byun, M.O., Park, S.C., 2008. Plant gene reponses to frequency-specific sound signals. Mol. Breed. 21, 217–226.

14. Johnson, K.A., Sistrunk, M.L., Polisensky, D.H., Braam, J., 1998. Arabidopsis thaliana responses to mechanical stimulation do not required ETR1 or EIN2. Plant Physiol. 116, 643–649.

15. Johnson, P.R., Ecker, J.R., 1998. The ethylene gas signal transduction pathway: a molecular perspective. Annu. Rev. Genet. 32, 227–254.

16. Jomdecha, C., Prateepasen, A., 2011. Effects of pulse ultrasonic irradiation on the lag phase of Saccharomyces cerevisiae growth. Lett. Appl. Microbiol. 52, 62–69.

17. Katz, E., Lagunes, P.M., Riov, J., Weiss, D., Goldschmidt, E.E., 2004. Molecular and physiological evidence suggests the existence of a system II-like pathway of ethylene production in non-climacteric citrus fruit. Planta 219, 243–252.

18. Kesanakurti, D., Kolattukudy, P.E., Kirti, P.B., 2012. Fruit-specific overexpression of wound-induced tap1 under E8 promoter in tomato confers resistance to fungal pathogens at ripening stage. Physiol. Plant. 146, 136–148.

19. Lelièvre, J.-M., Latché, Jones, A., Bouzayen, B., Pech, M., J-C, 1997. Ethylene and fruit ripening. Physiol. Plant. 101, 727–739.

20. Lin, Z.F., Hong, Y., Yin, M., Li, C., Zhang, K., Grierson, D., 2008. A tomato HD-Zip homeobox protein, LeHB-1, plays animportant role in floral organogenesis and ripening. Plant J. 55, 301–310.

21. Lin, Z.F., Zhong, S.L., Grierson, D., 2009. Recent advances in ethylene research. J. Exp. Bot. 60, 3311–3336.

22. Lincoln, J.E., Fischer, R.L., 1988. Diverse mechanisms for the regulation of ethyleneinducible gene expression. Mol. Gen. Genet. 212, 71–75.

23. Lincoln, J.E., Campbell, A.D., Oetiker, J., Rottmann, W.H., Oeller, P.W., Shen, N.F., Theologis, A., 1993. Le-Acs4, a fruit ripening and wound-induced 1-aminocyclopropane-1-carboxylate synthase gene of tomato (Lycopersicon esculentum)- expression in E. coli, structural

characterization, expression characteristics, and phylogenetic analysis. J. Biol. Chem. 268, 19422–19430.

24. Liu, M., Diretto, G., Pirrello, J., J-, P., Li, Z., Giuliano, G., Regad, F., Bouzayen, M., 2014. The chimeric repressor version of an ethylene response factor (ERF) family member, Sl-ERF.B3, shows contrasting effects on tomato fruit ripening. New Phytol. 203, 206–218.

25. Manavella, P.A., Arce, A.L., Dezar, C.A., Bitton, F., Renou, J.-P., Crespi, M., Chan, R.L., 2006. Cross-talk between ethylene and drought signaling pathways is mediated by the sunflower Hahb-4 transcription factor. Plant J. 48, 125–137.

26. Martel, C., Vrebalov, J., Tafelmeyer, P., Giovannoni, J.J., 2011. The tomato MADSBox transcription factor RIPENING INHIBITOR interacts with promoters involved in numerous ripening processes in a COLORLESS NONRIPENINGdependent manner. Plant Physiol. 157, 1568–1579.

27. Monshausen, G.B., Gilroy, S., 2009. Feeling green: mechanosensing in plants. Trends Cell Biol. 19, 228–235.

28. Pan, I.L., McQuinn, R., Giovannoni, J.J., Irish, V.F., 2010. Functional diversification of AGAMOUS lineage genes in regulating tomato flower and fruit development. J. Exp. Bot. 61, 1795–1806.

29. Takahashi, H., Suge, H., Kato, T., 1992. Growth promotion by vibration at 50 Hz in rice and cucumber seedlings. Plant Cell Physiol. 32, 729–732.

30. Vrebalov, J., Ruezinsky, D., Padmanabhan, V., White, R., Medrano, D., Drake, R., Schuch, W., Giovannoni, J., 2002. A MADS-box gene necessary for fruit ripening at the tomato ripening inhibitor (rin) locus. Science 296, 343–346.

31. Vrebalov, J., Pan, I.L., Arroyo, A.J.M., McQuinn, R., Chung, M., Poole, M., 2009. Fleshy fruit expansion and ripening are regulated by the tomato SHATTERPROOF gene TAGL1. Plant Cell 21, 3041–3062.

32. Weinberger, P., Measures, M., 1978. Effects of the intensity of audible sound on the growth and development of Rideau winter wheat. Can. J. Bot. 57, 1036–1039.

33. Yang, S.F., Hoffman, N.E., 1984. Ethylene biosynthesis and its regulation in higherplants. Annu. Rev. Plant Physiol. 35, 155–189.

34. Yokotani, N., Tamura, S., Nakano, R., Inaba, A., McGlasson, W.B., Kubo, Y., 2004. Comparison of ethylene-wound-induced responses in fruit of wild-type, rin and nor tomatoes. Postharvest Biol. Technol. 32, 247–252.

Chapter 4

Sugars in peach fruit: a breeding perspective

Marco Cirilli, Daniele Bassi & **Angelo Ciacciulli**

Department of Agricultural and Environmental Sciences (DISAA), University of Milan, via Celoria 2, 20133 Milan, Italy

The last decade has been characterized by a decrease in peach (*Prunus persica*) fruit consumption in many countries, foremost due to unsatisfactory quality. The sugar content is one of the most important quality traits perceived by consumers, and the development of novel peach cultivars with sugar-enhanced content is a primary objective of breeding programs to revert the market inertia. Nevertheless, the progress reachable through classical phenotypic selection is limited by the narrow genetic bases of peach breeding material and by the complex quantitative nature of the trait, which is deeply affected by environmental conditions and agronomical management. The development of molecular markers applicable in MAS or MAB has become an essential strategy to boost the selection efficiency. Despite the enormous advances in 'omics' sciences, providing powerful tools for plant genotyping, the identification of the genetic bases of sugar-related traits is hindered by the lack of adequate phenotyping methods that are able to address strong within-plant variability. This review provides an overview of the current knowledge of the metabolic pathways and physiological mechanisms regulating sugar accumulation in peach fruit, the main advances in phenotyping approaches and genetic background, and finally addressing new research priorities and prospective for breeders.

1. INTRODUCTION

Breeding programs of peach and nectarine (*Prunus persica* L.) have traditionally focused on the selection of traits associated with fruit appearance and textural attributes.1 Fruit size, color, and firmness represent crucial components for retailers, making peaches attractive for consumers, resistant to postharvest handling, and allowing an extended shelf-life.2 Despite the significant improvements by breeding of such characters and the large number of newly released cultivars, in recent decades, a general decrease of peach consumption in the USA and European countries is observed.3 Poor or inconsistent flavor quality has been often recognized as the major limiting factor for acceptance by the consumers, shifting preferences toward other more flavorful fruit types. The peach cultivars present on the marketplace are often judged flavorless and appear to lack the strong 'peach flavor' expected by consumers, even when harvested at optimum maturity stage.4,5The improvement of peach quality represents a crucial aspect for promoting consumption, prompting breeders toward the selection of novel and more flavorful cultivars6 that can develop flavors before the onset of the softening process. Many cultivars released in the last twenty years show important progress in flavor, attested by the introduction of the 'low-acid' (LA) trait, which characterizes cultivars, such as Big Top, which has a markedly sweet taste, due to both the low level of acidity and a medium to high sugar content.7 Although LA cultivars have been mainly developed for Asian market preferences, a general trend to increase peach sweetness has also been noted in important USA and European breeding programs.8,9 For example, among more than 300 cultivars released in the USA during the last decade, approximately 20% feature high soluble solids (more than 15° Brix), of which only 16% are early ripening, almost exclusively nectarines. 10Important international research projects are focusing on the development of new molecular tools to elucidate the genetic bases of important fruit quality traits, such as the projects RosBREED and FruitBreedomics.11,12

2. ROLE OF SUGARS IN PEACH QUALITY

For practical purposes, sugar content is estimated through the measure of the soluble solids concentration (SSC%, ° Brix). Although the sugar content is significantly correlated with SSC, theR^2 values are often variable, ranging from 0.33 to 0.72, depending on the contribution of other optically active compounds, such as pectins, salts and organic acids.[13,14,15] However, SSC is a reasonable surrogate measure of sugar content and the overall evaluation of peach fruit quality.[16]

Flavor is a combination of taste, aroma and mouth-feel attributes, defining a specific sensory profile that ultimately affects consumer judgment regarding the overall quality of fruits. In peach, taste largely depends on the water-soluble compounds, such as sugars and organic acids, conferring a sweetness and/or sourness sensation, and phenolic compounds, conferring astringency or bitterness.[17] Sugars represent a fundamental component of fruit edible quality, predominantly conferring sweetness, one of the main attributes influencing the degree of consumer satisfaction for peaches.[18,19] The sweetness intensity depends on the overall sugar amount as well as on the specific sugar profile (the relative content of each individual sugar) because fructose, glucose and sorbitol have approximately 1.7, 0.8 and 0.6-fold the sweetener power of sucrose.[20] The early findings of Robertson and Meredith[21] suggest that low-quality peaches contain lower fructose and higher sorbitol and glucose compared with high-quality peaches. However, the level of acidity strongly affects the sweetness perception. As demonstrated by the comparison of the chemical analysis and sensory profiles, sweetness is mainly correlated with the ratio between sugars and acids, the overall organic acids concentration, the amount of citrate and shikimate and fruit juiciness.[22,23,24,25,26] Sugars also affect mouth-feel attributes and aroma perception; in particular, sucrose and sorbitol are highly correlated with the overall taste and aroma.[24]

A series of early studies associated SSC levels with increased consumer acceptance or lack of flavor.[22,27] The degree of liking for peaches widely depends on the ripe soluble solid concentration (RSSC), which is the SSC of ready-to-eat peaches.[28,29] The EU has established a minimum of 8° Brix for the peach market (Commission Regulation No. 1861/2004), despite many authors suggesting the introduction of higher SSC values, depending on the acidity levels.[30,31,32] However, the establishment of a universal and reliable SSC value, assuring a minimal level of consumer satisfaction, is a difficult task. In general, consumer acceptance tends to increase along with RSSC levels, particularly when this parameter is low.[33] In normal acidic cultivars, consumer acceptance increases rapidly with RSSC until a threshold value, specific for each cultivar, is exceeded; then, a further increase does not produce significant changes. In LA cultivars, the relationship between consumer acceptance and RSSC is more linear, progressively increasing until approximately 100% satisfaction at maximum RSSC values.[28,29] In addition, the optimal sugar contents for peaches vary depending on the consumer group involved.[34] Many researchers highlighted the importance of cultivar grading based on their dominant organoleptic characteristic, then developing a reliable minimum quality index within each group, thus helping to match fruit characteristics to consumers' specific requirements.[35]

3. SUGAR ACCUMULATION DURING PEACH FRUIT DEVELOPMENT AND RIPENING

Peach fruit accumulates different types of soluble sugars and sugar alcohols, mainly sucrose (Suc), glucose (Glc), fructose (Fru) and sorbitol (Sor). Suc is the predominant sugar in the peach mesocarp at maturity, accounting for approximately 40 to 85% of the total sugars content, followed by Glc and Fru (in variable ratios), together representing approximately 10–25%, and Sor, accounting for less than 10%.[13,22,36,37] Mature peach fruit also contains detectable amounts of other sugars, such as maltose, isomaltose, raffinose, xylose, trehalose, 1-*O*-methyl-glucoside and fucose, and the polyols galactinol, glycerol, myo-inositol and maltitol.[38] Starch is accumulated at the early stages of fruit development and is then rapidly metabolized, becoming almost undetectable at harvest.[39] The total sugar content increases continuously during peach development up to full maturity, remaining constant or slightly decreasing during postharvest storage.[40] Hexoses are the most abundant sugars in immature peach fruit until the beginning of rapid growth by cell elongation (S-III), when Suc becomes the predominant type.[41, 42] An interesting exception has been reported for the cultivar 'Allgold', in which Suc is the main sugar, even at the early stages of fruit growth,[37] an aspect not further investigated in peaches. The accumulation pattern of Glc and Fru is generally characterized by reduced variations along fruit development, although their content is slightly higher at the early stages and progressively declines until maturity.[38,43,44] The Glc:Fru ratio varies depending on genotype, usually from 0.8 to 1 for most commercial cultivars and is approximately constant across fruit development stages.[45, 46] The Suc content slowly increases in young developing fruit, remaining low until the S-III stage, when the accumulation rate increases, reaching maximum content prior to the onset of ripening, during the stage of fruit growth slowdown (S-IV).[38] Sor levels remain low throughout fruit development, although peak accumulation has been detected during the S-II to S-III transition, followed by a sharp decline during the ripening process.[38,39] A similar peak at pit hardening was observed for other sugars, such as xylose, raffinose, maltose and galactinol in cv. 'Dixiland'.38 In particular, raffinose and galactinol are key antioxidant molecules related to stress protection, and their content is highly variable among peach varieties.47 During postharvest storage, Suc remains constant or slightly decreases, whereas hexoses increase or decrease,38,40 depending on the cultivar and the storage conditions.

4. SUGAR METABOLISM IN PEACH FRUIT

A considerable amount of research has been conducted on carbohydrate biosynthesis, transport and accumulation in peaches. However, functional validation of candidate genes is particularly complex due to the recalcitrance of peach to *in vitro* regeneration.48 Most of the knowledge at the functional level originates from other *Rosaceae* species or model plants, which provide useful information regarding the general aspects of carbohydrate metabolism, but do not allow for framing of the possible specific behavior in peaches.

Carbohydrate partitioning in peach fruit is regulated by a complex network of metabolic activities and physiological processes, including carbohydrate biosynthesis in source tissues, long-distance translocation by phloematic flux, metabolism and accumulation in fruit sink tissues. Sor and Suc are the main translocated assimilates in the phloem sap of the Rosaceae family.49 In mature peach leaves, Sor accounts for approximately 70% of the transported carbohydrates (approximately 2–3:1 ratio compared to sucrose, depending on cultivar), whereas Glc and Fru are present in lower amounts.50 The sugar unloading pathways in peach fruit tissues have not been fully elucidated nor has the primary carbohydrate source supplying the rapid fruit growth and Suc accumulation during the S-III stage. Evidence has been provided for an apoplastic Suc transfer at the S-I and S-III stages,42 which is also observed in apple and pear,51,52 and a preferential utilization of Suc in growing fruits.53 However, both apoplastic and symplastic routes have been proposed for sugar uptake54 because in other species, the unloading route varies according to the fruit developmental stage and growth conditions.55,56 Membrane sugar transporters regulate carbohydrate movement between compartments and may exert strict control on their fluxes. At the molecular level, peach genome availability has allowed for the identification of major sugar-transporter gene family members. The involvement of SOTs transporters in active Sor unloading pathways has been elucidated in sour cherry,57 and 10 *SOT* genes have been identified in peach, three of which are expressed in mesocarp tissues.58 Moreover, three membrane-localized Suc/H+ symporter proteins (SUTs) involved in apoplastic loading/unloading have also been characterized: *PpSUT1*, barely detectable in fruit tissues, *PpSUT2,* mainly expressed in phloematic cells and *PpSUT4*, the most abundant transcript, expressed in both parenchyma and phloem tissues.42 *PpSUT2* is predominantly located on the plasma membrane, suggesting a role in Suc retrieval from the symplastic continuum into the phloem. In the S-I stage, *PpSUT4* is localized in the tonoplast and has a role in the regulation of Suc release to sustain cell metabolism, whereas at the S-III stage, the expression of *PpSUT4* decreases, allowing for the start of Suc accumulation into the vacuole. In addition to sorbitol and sucrose transporters, 29 members of the hexose transporter gene families (STPs, TMTs and HTs) are also present in the peach genome, with at least 13 isoforms expressed in the mesocarp tissues.58,59,60 Recently, a novel family of conserved Suc transporters was discovered in Arabidopsis, the SWEET uniporter proteins, localized at the plasma-membrane or tonoplast.61,62 In peach, 15 putative SWEET coding sequences were found, two of which are expressed in the fruit, suggesting a possible conserved role in Suc distribution.42 Molecular and functional evidence are still missing, as for the rest of the hexose transporters families.

The sugar distribution within the cell compartments of the fruit mesocarp varies depending on the developmental stage, shifting from a predominant cytoplasmic localization in young fruit to the vacuole at maturity.63 Upon translocation, Suc is cleaved to Glc and Fru by sucrose synthase (SUS) and invertase (INV) enzymes.64 Both enzymes act on the Suc substrate, but SUS yields Fru and UDP-glucose (UDPG), enabling reversible Suc synthesis, whereas INV converts Suc irreversibly into Glc and Fru. A total of six *PpSUS* genes are present in the genome, all expressed in mesocarp tissues. *PpSUS1,* the only studied isoform, is prevalently expressed in fruit compared to leaves and is upregulated during fruit ripening and postharvest storage.65 However, the enzyme activity of SUS remains constant or only slightly increases throughout fruit ripening.44,65,66 Regarding the invertase gene family, six cell-wall (CWINV), two vacuolar (VINV) and eight neutral/alkaline (NINV) genes were identified, with at least three, two and seven isoforms expressed in the fruit mesocarp, respectively.58 CWINVs and VINVs are acid invertases (AI), localized to either the cell-wall or vacuole, whereas NINVs are neutral invertases (NI) with cytosolic localization.66 In model plants that import and metabolize sucrose alone, extracellular invertase plays a pivotal role in sugar partitioning by facilitating apoplastic phloem unloading.68,69,70 For example, a lack of acid invertase activity characterizes the sucrose-accumulating species *S. chmielewski*, and the silencing of the vacuolar invertase gene *TIV1* in tomato increases sucrose, thereby reducing the hexoses content.71,72 In peach, the physiological relevance of invertase activity has not be fully addressed. Both AI and NI activities strongly increase during the S-I to S-II transition, remaining stable until ripening and after harvest.38 In contrast, the capacities of NI and AI are stable throughout fruit development in a BC2 inter-specific [SD × Summergrand] × Zephyr population (S × Z).44 Expression studies of four neutral invertase isoforms (*PpNI1-4*) suggest a major role for *PpNI2*, whose transcript strongly increases at the S-III stage concomitantly with the rise in Suc accumulation, whereas *PpNI1* is mainly expressed during the early stages of peach development.38,73 Molecular evidence regarding acid invertases is limited to the *PpCWINV1* isoform, mainly expressed in leaves.66 As suggested in apple,74 and therefore also conceivable for sorbitol-transporting species, such as peach, Suc hydrolysis by invertase may play only a minor role in fruit sugar assimilation.

In fruit, Sor can be rapidly oxidized into Fru by NAD-dependent sorbitol dehydrogenase (SDH) or sorbitol oxidase (SO).39 Despite SO activity being detected in peach fruit,75 it plays only a minor role in Sor metabolism.76 SDH is encoded by seven genes in peach, with at least four detected in fruit.58 The expression pattern of one *PpSDH* gene during peach development showed two peaks, at the end of the S-I stage and during fruit ripening, which is in agreement with the rates of SDH activity.38,39,77 Comparing the commercial and native peach accessions showing high and low Fru content, respectively, Kanayama *et al.*78 suggested that the metabolic capacity of SDH might be responsible for the different Fru levels in fruits. However, this hypothesis was not confirmed because no significant difference in SDH activity was observed in progeny segregating for Fru content.44

The pool of Glc and Fru resulting from sucrose and sorbitol metabolism can be stored in the vacuoles or phosphorylated to glucose-6-phosphate (G6P) and fructose-6-phosphate (F6P) by hexokinase (HK) and fructokinase (FRK), whereas the interconversion between F6P, G6P, UDPG and glucose-1-phosphate (G1P) is catalyzed by phosphoglucoisomerase (PGI), phosphoglucomutase (PGM) and UDP-glucose pyrophosphorylase (UGP) by readily reversible reactions.79 G1P enters the starch biosynthesis pathway, whereas F6P can enter the glycolytic pathway or can be combined with UDPG to synthesize Suc by sucrose phosphate synthase (SPS) and sucrose-phosphate phosphatase (SPP) through an essentially irreversible reaction.[80,81] The enzymes PGM and UGP, which are linked to hexose phosphates metabolism, displayed the highest capacities,[44] although the low G1P and starch contents in peach fruit suggest that UGPase activity may be directed toward UDPG biosynthesis. HK and FRK capacities are higher during the final stage of peach fruit development in genotypes with high Glc:Fru ratios.[44] Among the four putative*SPS* genes identified in peach, only *PpSPS1* and *2* have been studied in detail. *PpSPS1* did not significantly change during peach fruit development, whereas *PpSPS2* was upregulated during the S-III to S-IV transition, with a further increase detected postharvest.[38,66] In contrast, SPS activity increased only during the early stages of fruit development.[44]

Collectively, the wide number of studies analyzing the gene expression and activities of key structural enzymes does not provide a clear framework that explains sugar accumulation patterns in peach fruit. As demonstrated by extensive analysis of sugar-related enzymes,[44] temporal variations in enzymatic activities are too small compared to the broad changes in metabolite concentrations, revealing a remarkable stability across years and genotypes with variable sugar compositions. Genome-wide data on the transcriptional profiling of both sugar transporters and structural genes along peach fruit development are not available; thus, the role of different isoforms belonging to multigenic families is unclear.

5. PHENOTYPIC AND GENETIC VARIABILITY OF THE SUGAR CONTENT IN THE PEACH GERMPLASM AND BREEDING POPULATIONS

In many crops, wild accessions have been proven to be a useful source for the introgression of important traits associated with fruit quality and/or for the mapping of the genomic regions undergoing strong selective pressure during domestication. This aspect could be of particular importance for *P. persica*, characterized by a narrow genetic base and low intraspecific diversity.[82,83] The comparison of fruit composition among peach cultivars, *P. davidiana* (the wild-related species more distant to peach) and hybrids, suggests that peach domestication and hybridization have resulted in a large increase of sugar content.[84] In *P. davidiana,* the fruit sugar content is significantly lower than peach (<100 mg/g of dry weight), Suc is predominant and Fru is higher than Glc. Information regarding the variability of the sugar composition in F1 peach × *P. davidiana* hybrids is limited to a few Summergrand × P1908 selections (SD) showing a total sugar content similar to peach and higher Glc and Fru content with respect to peach parentals.[84] In BC2 interspecific [SD × Summergrand] × Zephyr (S × Z) progenies, the Glc:Fru ratio is approximately 1:1 in most seedlings, despite approximately 10–15% of the individuals showing transgressive segregation for high Glc content compared to peach parentals.[85,86] In other wild species, *P. davidiana var. potaninii, P. mira, P. kansuensis* and *P. ferganensis,* the fruit sugar composition remains largely uncharacterized, as with the rest of the Eastern Asia germplasm. Recently, comparative population genomic studies have begun to explore the genetic variability associated with landraces, ornamental and wild species belonging to the Chinese peach germplasm, providing rough information regarding the fruit SSC, ranging from 7.7° to 17.5° in *P. persica* and 10° to 16° Brix in wild accessions (Supplemental File 1).[87,88,89] The main commercial Chinese peach cultivars show lower inbreeding levels and a greater genetic diversity compared to Western cultivars (notably from the USA, and European and USA-derived).[90]

The sugar composition of many commercial, native and ornamental peach cultivars of Japanese germplasm was investigated by Moriguchi *et al.*[36] Suc was the predominant sugar in native and ornamental peaches, with similar amounts compared to commercial peaches. Moreover, Japanese and Western cultivars were characterized by nearly equal percentages of Fru and Glc, contrary to native and ornamental peaches, which are characterized by a high Glc:Fru ratio and Sor amounts (Supplemental File 1). The majority of works regarding sugar composition and SSC variability have been conducted on European and USA cultivars (Supplemental File 1).[13,21,22,90] Detailed information is available for a limited number of cultivars, making it difficult to unbundle the environmental effects and, thus, provide a more complete overview of intra- and intercultivar variability. Reig *et al.*[9] evaluated the SSC and sugar composition of 108 recently developed peach cultivars from public and private USA and European breeding programs, reporting a wide range of variability for SSC (9.5°–19.8° Brix) and total sugar content (89.1–184.5 gL^{-1}). Font i Forcada *et al.*[91] comparing 94 traditional Spanish (mostly non-melting flesh types) and worldwide accessions, reported extensive variability of both the overall sugar content and composition. The total sugar content varied from 63 to 136 g/kg FW (Suc 35–98, Glc 4–15, Fru 2–14, and Sor 2–35), whereas SSC ranged from 12° to 18° Brix. However, in a survey of 120 Eastern and Western cultivars released during the last century (mainly from Japan, Korea and the USA), reduced variability in fruit SSC was reported, ranging from 9.1° to 13.9° Brix, with significant but very low differences between

Eastern and Western cultivars (overall average values of 11.7° and 10.8°, respectively).[92]

The variability of the sugar contents and their profile has been evaluated in many intraspecific and interspecific progenies (Table 1).[1,14,37,44,85,93,94,95] The analysis of traits of interest in segregating populations allows estimation of the crucial parameters for the establishment of an effective breeding strategy, such as environmental effects, phenotypic frequencies and heritability. Overall, seedlings from different cross populations exhibited wide phenotypic variability for the total sugar content (from 60 to almost 140 mg/g FW) and for individual sugars and a broad range of SSC (from below 9 to over 17° Brix), suggesting that there is genetic potential to improve the sugar content and composition in commercial peaches. All traits displayed continuous variation, following a normal or bimodal distribution among progenies, which is typical of quantitative or polygenic inheritance. Depending on the season, phenotypes can significantly deviate from these distributions. Suc was the predominant sugar in almost all progenies, whereas Glc and Fru ranged within 0.4–2.5. Sor was also variable within progenies, although the range of variability rarely exceeded the Glc or Fru amount. Although the mean values of each sugar were variable in the progenies within the parental range, a remarkable percentage of transgressive individuals with higher content with respect to either parental has been reported in almost all populations. Such transgressive segregation indicates that it could be possible to select for high sugar content in most seedling populations. Some works provided a rough estimate of sugar heritability.[37,96] A moderate to high broad sense heritability has been calculated for SSC (0.33 to 0.72), based on 13 genotypes and approximately 2000 observations, and low heritability for both total and individual sugar (approximately 0.20, except for Sor, 0.50).[37,96] In contrast, high heritability of Suc, Glc, Fru, Sor and total sugars, ranging between 0.65 and 0.90, was estimated in a reciprocal cross 'Zaoxing' (an LA flat peach) × 'Zaolupan' (a non-LA round peach), despite the size of the progenies being insufficient to provide a reliable evaluation.[46] Interestingly, Wu *et al.*[46] demonstrated that maternal inheritance did not affect the sugar composition.

Table 1. Mean values and range of variability for the soluble solid content (SSC), total and individual sugars, detected in fruits of the most relevant analyzed cross populations available in literature

Population	n	SSC (°Brix)		Total Sugars (mg/g FW)		Sucrose (mg/g FW)		Glucose (mg/g FW)		Fructose (mg/g FW)		Sorbitol (mg/g FW)		Ref
		Mean	Min–Max	Mean	Min–Max	Mean	Min–Max	Mean	Min–Max	Mean	Min–Max	Mean	Min–Max	
F1 'Venus' x 'Big Top'	75	13.9 ± 0.2	11.2–17.5	89.7 ± 1.6	67.4–138.9	58.4 ± 1.2	40.7–102.3	12.2 ± 0.3	8.3–23.4	12.4 ± 0.2	8.9–19.1	6.6 ± 0.5	1.7–19.5	95
F1 'Andross' x 'Calante'	19	12.2	7.6–17.5	85.9	36.0–109.4	64	28.2–84.4	6.6	2.3–14.6	10.3	3.8–16.1	5.1	0.9–10.6	14
F1 'Andross' x 'Crown Princess'	9	11.0		67		51.6		6.4		7.6		1.4		
F1 'Andross' x 'Rich Lady'	9	11.1		73.3		54.9		5.8		9.0		3.6		
F1 'Andross' x 'VAC-9511'	6	11.9		67		49.6		6.7		9.1		1.6		
F1 'Babygold-9' x 'Crown Princess'	19	9.7		69.4		53.5		5.7		7.9		2.4		
F1 'Babygold-9' x 'VAC-9510'	15	11.0		67.7		52.5		5.7		7.4		2.2		
F1 'O'Henry' x 'VAC-9514'	14	13.2		77.7		54.3		8.0		10.1		5.3		
F1 'O'Henry' x 'VAC-9515'	8	13.5		80.3		60.9		6.8		9.3		3.2		
F1 'O'Henry' x 'VAC-9516'	11	12.5		75.5		58.2		6.5		6.9		3.9		
F1 'Orion' x 'VAC-9513'	3	11.6		67.2		47.1		7.3		8.9		3.9		
F1 'Red Top' x 'VAC-9513'	37	9.9		64.0		48.4		5.6		7.8		2.1		
F1 'Rich Lady' x 'VAC-9511'	8	12.3		67.6		50.0		6.7		8.4		2.5		
F1 'VAC-9512' x 'VAC-9511'	10	11.6		77.3		57.1		7.0		9.0		4.2		
F1 'VAC-9520' x 'VAC-9517'	37	12.2		75.6		56.5		7.3		8.9		2.8		
F1 'Ferjalou Jalousia' x 'Fantasia'	nr	12.2 ± 0.4	nr	nr	nr	72.9 ± 2.5	nr	4.9 ± 0.4	nr	7.0 ± 0.3	nr	7.0 ± 0.3	nr	93
F2 'Ferjalou Jalousia' x 'Fantasia'	63	11.45	8-14	nr	nr	67.85	40-90	10.2	5–15	10.8	6–17	1.2	0.5-6	94
BC2 'Summergrand' x 'Zephir'	140	nr	10.1–19.3	92 ± 18	56-167	nr	20–150	nr	5–35	nr	2.5–25	nr	2.5–20	86, 120
F1 'Zaolupan' x 'Zaoxing'	46	nr	nr	96 ± 15	69-120	66 ± 17	34–110	13 ± 4.3	6.0–21	13 ± 3.6	6.8–19	4.3 ± 2.9	0.7–13	155
	43			86 ± 19	50-124	48 ± 16	19-81	16 ± 6.1	7.0–30	16 ± 6.9	6.7–34	6.3 ± 3.0	1.0–12	
F1 'Zaoxing' x 'Zaolupan'	89			91 ± 16	58-136	65 ± 18	30–120	12 ± 3.5	5.8–21	11 ± 2.8	5.2–19	2.8 ± 1.7	0.5–10	
	90			84 ± 18	48-124	47 ± 17	22-81	16 ± 6.1	7.1–28	16 ± 5.7	9.2–30	5.4 ± 3.4	0.6–18	

6. FIELD AND ENVIRONMENTAL FACTORS AFFECTING THE SUGAR CONTENT AND PROFILE

A large proportion of phenotypic variability for the peach sugar content depends on environmental factors and genotype-by-environment interactions. Fruit sugar variations in different growing seasons and locations, within trees of the same orchards, within the same tree and also within the fruit itself, are not negligible in comparison with the variation between genotypes.[97] As described before, the sugar content varies depending on the fruit developmental stage and is mainly regulated by carbohydrate supply, metabolic transformation and dilution effect due to variation in fruit volume.[98] These physiological and metabolic processes are influenced by field practices, such as irrigation,[99,100] fertilization,[101,102] rootstock:scion interactions,[103,104,105] training system,[106] pruning and canopy management.[107,108] Additionally, environmental variables, such as temperature, solar radiation, photoperiod, precipitation and soil patterns, influence the tree-growing environment and have been widely proven to result in wide variations in sugar accumulation.[16] The variability of the sugar contents between trees is smaller compared to the within-tree variability, which reaches differences up to 10° Brix.[109] Sources of within-tree variability have been mainly ascribed to fruit position in the tree, microclimatic gradients inside the canopy, leaf to fruit ratio and the vigor of fruit-bearing shoots. These factors mainly affect the availability of carbohydrates supporting fruit growth.[110] Nevertheless, even under condition of unlimited carbohydrate availability, as in thinned trees, differences in fruit sink activity result in substantial variations in fruit sugars.[111,112] Not negligible, the SSC variability within a single fruit can be as high as 4° Brix, with differences commonly found between the blossom/stem-end and between cheeks/suture.[113,114] For these reasons, the accuracy of the SSC and sugar content estimation could be improved by evaluating a larger number of fruits within a single tree rather than by increasing the number of trees *per*genotype.[85]

Climate and crop load are considered the most important sources of year-to-year variability.[115] The relationship between yield and SSC is generally negative, although the degree of correlation varies depending on genotypes.[1,14] High crop load has been negatively associated with carbohydrate accumulation in fruits because it alters the source–sink balance, causing an increase of sink competition among fruits.[116] Several authors have reported contradictory results regarding the seasonal variability of the sugar composition.[1,37,85,86]

The effects of environmental conditions and field practices on sugar accumulation in peaches have been deeply investigated, helping to explain the major causes of within tree variability in fruit composition. Collectively, the authors agree that the SSC, total and individual sugar contents are strongly affected by seasonal variability, in contrast to the sugar profile, which is relatively constant across environments and genotypes. However, how the environment affects the metabolic fluxes at both the enzymatic and gene expression levels remains to be elucidated.

7. MODELING SUGAR ACCUMULATION IN PEACH FRUIT

Sugar accumulation in fruit is a complex quantitative trait, affected by environmental conditions and dependent on many interconnected physiological and metabolic processes, and controlled by multiple genes that interact with the environment and crop management. Quantitative trait loci (QTL) underlying the traits of quantitative nature often explain low phenotypic variability, with reduced stability across years and strong genotype-by-environment or genotype-by-management interactions. The virtual profiling of phenotypes through ecophysiological process-based simulation models (PBSM) represents a promising approach to overcome such difficulties due to their ability to mimic complex systems and to integrate multiscale levels of control.117,118,119 Ecophysiological modeling relies on the analysis of parameters involved in the development of traits, instead on their direct measurement.120 The identification of key genetic parameters allows for simulation of the performance of genotypes in many environments because they are independent of the environment. Such parameters can be considered as genotypic traits and are more suitable for association studies compared to the direct measure of complex traits affected by the environment. In peach, the SUGAR model has been developed to simulate the variations of the sugar composition during the S-III stage of fruit development based on carbohydrate supply, changes in fruit metabolism and assimilates dilution.76,98 Simplified forms of the model have also been applied to simulate the refractometric index,121 to analyze genotypic variation of the total sugars content in a segregating population,122 or to account for different Glc:Fru ratios in some genotypes.46 The SUGAR model was implemented in a more general 'virtual peach fruit model', integrating submodels for carbon assimilation, allocation and water fluxes to predict the evolution of important fruit quality traits.123 Among the several genotypic parameters that compose the model, 10 were able to explain 52% of variability in the fruit sweetness index.124 In particular, hydraulic conductance of the fruit surface (*aL*) and the coefficient of the transfer function between sugars and other compounds (k_{sugar}) are the main parameters linked to sugar content and also highly correlated with other associated with fruit growth demand and duration.122 Some QTLs associated with ecophysiological parameters and influencing fruit sugar metabolism, although showing a low stability across years, were identified.125 For example, r_{su} (the Suc proportion of the total sugar amount) co-located on linkage group (LG) 1 with QTLs for fresh mass and early fruit growth. QTLs for sugar content on LG2, 4, and 7, co-located with QTLs for parameters involved in fruit water fluxes, such as *aL* and ρ (permeation coefficient of the fruit surface to water vapor). QTLs for growth duration (dd_{max}) were co-located with those for *aL*, k_{sugar} and r_{su}. This preliminary study demonstrated the potential usefulness of ecophysiological models. However, to increase their applicability, the models should also include additional genetic information via genotype-dependent parameters by easily measurable physiological traits and known QTLs or, even better, genes.126

8. QTL-MAPPING OF SUGAR-RELATED TRAITS IN PEACH

The improvement of peach quality traits has been achieved through traditional phenotypic selection within the seedling populations. However, this procedure is expensive and time-consuming, requires the screening of a large number of individuals, and above all, is mostly effective in fixing highly heritable traits. The selection for increased SSC, reportedly having moderate heritability, has allowed for a certain improvement of the sugar content in medium to late ripening cultivars, despite the variations caused by environmental and field practices. The development of peach cultivars with short fruit development periods (FDP) with high SSC is difficult due to the negatively correlated selection response between FDP and SSC,96 although current USA breeding programs suggest the possibility to develop early-ripening peaches with moderately high SSC.10 In contrast, the improvement of traits related to individual sugar contents, characterized by low heritability, is more complex to achieve by conventional selection. The identification of genetic determinants controlling these traits and the development of markers closely linked to relevant QTLs might allow for a marker-assisted selection (MAS) approach by pyramiding combinations of genes and assembling target traits more precisely.

QTLs for sugar-related traits have been mapped in a large number of species, including strawberry,127 apple,128 sweet cherry129 and apricot,130 indicating some levels of synteny among*Rosaceae* species. Despite the huge amount of available data, the genetic determinants underlining these QTLs have not been identified, except in a few cases for tomato, a model species for unveiling the molecular mechanisms regulating sugar accumulation in fruit. Among the identified loci in tomato, the most important are linked to altered sugar transport, starch biosynthesis or Suc invertase activities, such as the fine-mapped Brix 9-2-5 locus, which located an apoplastic invertase.69,131,132 A quantitative trait nucleotides (QTN) caused by an amino acidic substitution in the LIN5 gene is responsible for the altered activity between the cultivated and wild species allozymes.133

In peach, genetic determinants for some important fruit quality traits have been identified, excluding sugar-related traits, for which knowledge is still limited.134 QTL mapping experiments for the sugar content in peach have been conducted using biparental crosses for a limited number of genetic markers. The most significant QTLs associated with sugar-related traits currently identified in peach are shown in Table 2 and Figure 1. Early QTL mapping of an F2 progeny from a cross 'Bailey' × 'Suncrest' identified five loci associated with hexose concentrations, although the molecular markers could not be integrated in the *Prunus* reference map; therefore, information on QTLs locations cannot be used for a comparative mapping.135 A major QTL for the total soluble solid content was stably mapped on the central part of LG4, near the MD locus, in 'Ferjalou Jalousia' × 'Fantasia' (J × F),136 'Contender' × 'Ambra'(C × A),137 and S × Z progenies.86 Other QTLs were also detected on LG5 and 6, although in different map intervals.86,136 In sour cherry, a QTL for SSC was mapped on LG6, suggesting that this linkage group might be conserved between the two *Prunus* species.129 Additional loci for SSC were identified in a BC1 population derived from a cross between a peach selection 'IF7310828' and *P. ferganensis* 'P72' (IF × P72) at the top of LG6 and on the distal part of LG2,138 as also observed in a 'Bolero' × 'Oro A' population.137Recently, an association study involving a large number of accessions belonging to the USA reference set for the peach germplasm have identified two genomic regions linked to SSC, on the central part of LG6 and at the bottom of LG7, although unstable across the two years of phenotypic observation.139 Many of the QTLs for SSC have been confirmed by a large association study involving over 1000 accessions within the FruitBreedomics project, albeit apart from QTLs located near the MD locus, the others have intensities close to the threshold probability (Aranzana*et al.*, 2015, unpubl. data).

Table 2. Table 2 Localization and approximate genome position of the principal QTLs for the soluble solid content (SSC), total and individual sugars identified in peach. The statistical significance of the linkage between a QTL and marker is indicated according to the detection methods: *LOD score; ***P*-value; ***2 ln (Bayesian Factor)

Trait	LG	Statistical significance	Nearest marker	Locus	Approximate genome position	Population	Ref
SSC	2	3.3*	Pij1–AG20		Pp02:19 108 809..21 093 830	BC_1 IF × P72	136
	4	0.0023**	CC129 (pchgms2)		Pp04:2 086 577	BC_2 S × Z	86
	4	4.9*	BPPCT015	MD	Pp04:12 558 026..13 520 063	F_2 J × F	134
	4	47.8–49.4*	MD	MD	Pp04:11 138 518..11 140 641	F_2 C × A	135
	5	2.9*	MA026	D	Pp05:145 584..3 741 013	F_2 J × F	134
	5	0.00027**	AG46-AG108	G	Pp05:14 652 809..18 294 595	BC_2 S × Z	86
	6	2.9*	UDP98-416		Pp06:4 375 577	BC_1 IF × P72	136
	6	4.47***	ss_629062–ss_630302		Pp06:7 918 349..12 571 791	GWAS	137
	6	3.3*	UDP98-407		Pp06:20 450 677..21 030 866	F_2 J × F	134
	6	2.3*	MA014	S	Pp06:27 186 773	F_2 J × F	134
	7	7.80***	ss_708371–ss_752524		Pp07:1 125 816..8 336 521	GWAS	137
Fructose	1	$2.9e^{-15}$**	PC102	FRU	Pp01:9 959 357..12 857 908	BC_2 S × Z	86
	4	0.0072**	UDP96-003		Pp04:8 768 343	BC_2 S × Z	86
	4	10.5*	BPPCT015	MD	Pp04:10 497 063..14 742 215	F_2 J × F	134
	5	3.0*	MA026A	D	Pp05:145 584..3 741 013	F_2 J × F	134
	6	7.1*	CPPCT023–pchcms5		Pp06:20 450 677..21 030 866	F_2 J × F; T × E	134,138
	7	0.0008**	pchcms2		Pp07:18 688 565	BC_2 S × Z	86
Glucose	4	9.4*	BPPCT015	MD	Pp04:10 497 063..14 742 315	F_2 J × F	134
	5	3.4*	MA026A	D	Pp05:145 584..3 741 013	F_2 J × F	134
	6	7.9*	CPPCT023–pchcms5		Pp06:20 450 677..21 030 866	F_2 J × F; T × E	134,138
	7	0.0045**	pchcms2		Pp07:18 688 565	BC_2 S × Z	86
Sorbitol	4	0.0016**	UDP96-003		Pp04:8 768 343	BC_2 S × Z	86
	4	6.6*	BPPCT015	MD	Pp04:10 497 063..14 742 315	F_2 J × F	134
	5	2.2*	MA026A	D	Pp05:145 584..3 741 013	F_2 J × F	134
	6	21.9*	CPPCT023–pchcms5		Pp06:20 450 677..21 030 866	F_2 J × F; T × E	134,138
Sucrose	5	15.3*	MA026A	D	Pp05:145 584..3 741 013	F_2 J × F	134
	7	2.4*				F_2 J × F	134
	7	0.0088**	pchcms2		Pp07:18 688 565	BC_2 S × Z	86
Total sugars	4	3.2*	BPPCT015	MD	Pp04:10 497 063..14 742 315	F_2 J × F	134
	5	7.3*	MA026A	D	Pp05:145 584..3 741 013	F_2 J × F	134
	6	3.3*	UDP98-407		Pp06:17 642 515	F_2 J × F	134

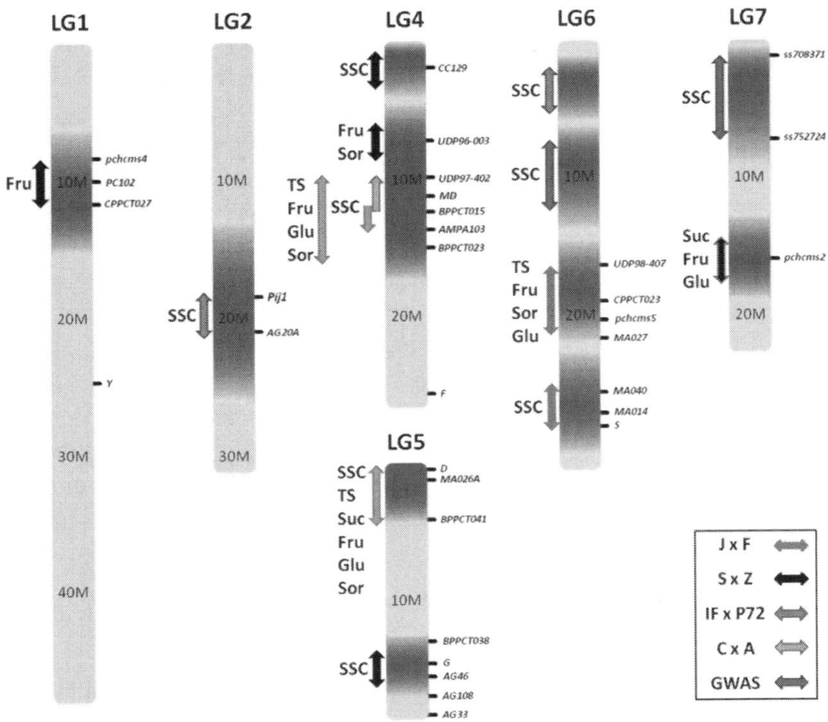

Figure 1. Graphical map of the position on linkage groups of the main sugar-related QTLs.

Main QTL mapping experiments for the content of individual sugars have been performed on J × F and S × Z populations, identifying several genomic regions.[86,93,94,136] However, only a few loci were confirmed across the two populations. A QTL with a main effect on the Suc content was mapped on LG5, in the region of the D locus in J × F progenies, and on the distal part of LG7 in S × Z progenies.[86,136] QTLs linked to Glc, Fru and Sor content all seem gathered in clusters on LG4, 5 and 6. However, in the S × Z population, a major QTL for Fru content was also found on LG1 (at the FRU locus) in addition to minor QTLs affecting both Glc and Fru co-located on the distal part of LG2 and 7.[86] Non-significant QTLs were identified for the total sugar content, except two minor QTLs located on LG5, at the D and G loci.[86,136] Unexpectedly, Quilot *et al.*[86] reported the presence of several favorable alleles for sugar content provided by *P. davidiana* parental, despite the low-sugar phenotype of its fruit. This aspect should be confirmed in the future, considering that the introgression of such valuable alleles in peach would require several rounds of backcrosses to eliminate the many unfavorable traits conferred by *P. Davidiana,* such as size, flavor and external appearance.

Overall, the mapping experiments suggest that both the total and individual sugar contents in peach fruit are governed by several QTLs with minor effects, often gathered in clusters. In many cases, such QTLs are unstable, due to strong environmental effects and are characterized by low LOD scores and small percentages of explained phenotypic variability. The number of loci governing sugar content is only roughly estimated, as well their genomic position, spanning regions on the order of 5–10 cM. Despite the availability of genome sequences and the attempts to identify candidate genes for the control of sugar-related traits,140 such intervals are too large to enable QTLs map-based cloning. The likelihood of hundreds of genes being present within these regions makes it difficult to identify the linked gene(s).

9. RELATIONSHIP OF THE SUGAR CONTENT WITH THE AGRONOMICAL AND POMOLOGICAL TRAITS

The evaluation of possible correlations between the sugar composition and physicochemical parameters is of particular importance for breeders, either to improve related traits simultaneously or to reduce undesirable side effects when selecting for one of the correlated traits. The relationships of sugar with fruit size, shape, flesh color, flesh adhesion to pit, pubescence and maturity date have been investigated.

The duration of the fruit development period (FDP) and maturity date (MD) affect the sugar contents, as broadly indicated by the reduced SSC of early-ripening fruits in comparison with mid-to-late ripening ones. A good correlation was reported between SSC and FDP (>0.60), and there was no relationship with blooming date.96 A significant and positive correlation between FDP and the SSC, Glc and Fru contents has been observed in a J × F progeny94 and in a germplasm collection showing extensive difference for maturity date.91 In contrast, these correlations were not evident in other progenies, although a general tendency towards higher SSC was noted in the latest ripening accessions. The correlation between ripening date and sugar content is supported by QTL co-localization on LG2136 and by the co-localization of a QTL for SSC at the MD locus, supporting the hypothesis of a pleiotropic effect.137 An NAC gene, a candidate for the control of ripening time in peaches, has been recently fine-mapped at the MD locus;141 however, it is unknown how it may affect sugar accumulation at the functional level.

The relationship between fruit weight (or size) and sugar content has obvious implications for fruit quality, although the relationship is complex and often not trivial. Some studies have documented a weak or not significant correlation between sugars and both fruit dry (DW) and fresh (FW) weight.91,96,142 For example, fruit fresh mass at harvest explained only a small amount of the total sugars variability in 14 different peach progenies.14 A significant relationship between the mesocarp DW, SSC and sugars was found in an S × Z population,86 and a negative correlation has been reported between the Glc content and both FW and DW.85 Although a correlation between fruit DW and sugar content is expected because carbohydrates compose approximately 60% of the DW, significant variability for sugar content has been found among fruits of similar weight because, as observed in tomato introgression lines, the proportion of DW consisting of sugars may vary depending on the metabolic utilization of the assimilates for growth or for storage.143 Differences in sink efficiency of individual fruits play a critical role in both DW and soluble sugar accumulation in peach.144 It has been hypothesized that assimilate partitioning within individual fruits depends on differences in the mesocarp cell number and size because their variability explains 77% of the variation in the Suc content between fruits within a single tree.145 A strong correlation between the sugar concentration and the rate of cell expansion has also been observed in peach,146 and in melon fruit, a preferential accumulation of Suc in larger cells has been demonstrated.147 Moreover, fruit cell expansion is also regulated by the water balance at the whole plant level and fruit transpiration, ultimately affecting sugar concentration via dilution.148,149 Crop load can significantly alter the relationship between fruit mass and SSC, particularly in high-load conditions.150 The correlation between fruit size and sugar content is evident when comparing peach and nectarine fruit. For example, the comparison of two peaches, 'Tropic Beauty' and 'Fla.M3-1', and their respective nectarine mutants 'TBN' and 'M3-1N', suggested that the increased sugar concentration in nectarines depends on a decrease of the fruit size compared to the original peach. Therefore, the mutation affects the fruit growth and not its ability to accumulate sugars.151 A higher SSC content was also observed in sibling peach and nectarine seedlings from three hybrid families,152as well in the nectarine mutant 'Yuval' compared to its original peach 'Oded'.153 The G locus, controlling the peach/nectarine trait, has been recently characterized by a variant discovery approach, identifying an LTR retro-element insertion in exon 3 of the candidate gene *PpMYB25* as the cause of the recessive glabrous phenotype.154 The fruit weight reduction in nectarines suggests a pleiotropic effect of this mutation on the sugar content. Alternatively, the reduced size depends on a tightly linked QTL. Nevertheless, contrasting results have been reported by mapping experiments with a QTL for fruit size and sugar content co-localized near the G locus in S × Z but not in J × F populations.86,93 In addition, QTLs for high Fru and Glc concentrations co-located with QTLs, showing a negative effect on fruit size on LG4, 5 and 7.86 The co-localization of QTLs with opposite allelic effects for sugar content and fruit weight has also been observed in tomato, suggesting that some genes may have pleiotropic effects.155

A general tendency for a greater sugar content in flat-peach has been observed by exploring germplasm resources.156 In progenies segregating for peach/nectarines and flat/round traits, SSC tends to decrease in round peach fruit compared to flat-peach and flat-nectarines, respectively.157As described above for locus G, the increase of SSC may be due to the pleiotropic effects of size reduction because locus S, controlling the flat-shape trait, is associated with a QTL for fruit weight on LG6 in a J × F progeny.93

A significant association between the sugar composition and flesh color has not been highlighted, although sugars tend to be higher in white-flesh cultivars.14,158,159 The co-localization on LG1 of the FRU locus, controlling the Fru content and sweetness,86 with the Y locus, controlling the yellow/white color of the flesh,160 might explain the association of both traits.

In the case of the freestone/clingstone trait (*F* locus), slightly but significantly higher SSC, total and individual sugar contents were found in freestone fruits.14 However, other authors reported no significant association,85 or, in contrast, higher total sugars in clingstone cultivars.91 Although a QTL for sugar content has been reported on LG4, where the F locus is located,86 the difference does not depend on the linkage of the two traits but rather on the genetic background of the analyzed genotypes.

10. CONCLUSIONS AND PERSPECTIVES

Presently, the development of new cultivars showing improved sensory attributes is a crucial task for breeders to raise the peach appeal on the fresh market. The improvement of eating quality of peach can be achieved by enhancing the sugar content or sweetness perception (for the low acid types), which are generally prized by consumers. However, as demonstrated by several studies, the increase of the sugars content is not always accompanied by a parallel increase in the liking degree. Therefore, the selection of new cultivars should be defined as a more specific target to be achieved, based on detailed information regarding consumer preferences and perceptions.

An improvement of the sugar content has been observed in recently released cultivars from different conventional breeding programs. Nevertheless, further and substantial improvements by classical breeding methods are limited by the scarce knowledge of the gene network expressing the phenotype. Although the strong influence of environmental factors and genotypic × environment interactions could mask the effects of minor QTLs, the genetic determinants regulating sugar accumulation could be identified, allowing for the identification of molecular markers to be used in marker-assisted selection (MAS) or breeding (MAB). Even when several minor QTLs are made available, marker-assisted approaches could represent the best strategy for an efficient screening procedure, through QTLs pyramiding. Unfortunately, none of the putative identified QTLs governing sugar-related traits in peach have been fine-mapped. The lack of detailed information on their effects and genome location, as well as their validation in different genetic backgrounds, prevents the development of suitable molecular markers for breeding purposes.

The availability and the high level of synteny among several *Prunus* genomes, together with a wide array of new technologies, offers unheard-of opportunities to breeders. NGS technologies have been implemented in linkage mapping experiments to develop novel and promising approaches, such as RAD-Seq,161 GBS,162 GS,163 and QTL-Seq.164 These powerful approaches could allow for the creation of high-resolution maps with a high number of molecular markers. However, these tools could suffer the same limitations as classic linkage mapping strategies because the increment of markers density is often not sufficient for unambiguous QTLs identification. Map resolution can be limited by the few recombinant events occurring in early generations and more often by the small size of the segregating progenies. The increase of progeny size is often difficult to achieve due to the time and the costs of field management. Alternative strategies of QTL mapping based on advanced backcross, NILs, RILs or double haploids, are not easy to perform in a tree crop. Furthermore, genome-wide association studies (GWAS) are ineffective in peaches, at least when applied to the European and USA germplasm, which are characterized by narrow genetic bases and high levels of linkage disequilibrium (LD).165

To exploit the full potential of the above described technologies and to increase the genetic gains through so-called precision breeding,166 the availability of accurate and high-throughput phenotyping tools is of paramount importance. Although data collection of sugar contents is not difficult from a technical standpoint, the phenotyping is hindered by the strong environmental effects and the high within-tree variability, in turn representing another reason for the lack of accuracy of the QTLs mapping experiments. A full understanding of the processes underlying the interconnections among yield, fruit growth, dry and fresh mass accumulation and sugars partitioning is essential in the environmental variability and its interaction with genotype. Ecophysiological modeling represents a viable approach to improve the phenotyping of complex traits; however, such models are currently in progress and are rarely applied by breeders because they depend on parameters that are not easily measurable.

In conclusion, the following priorities should be followed for the improvement of sugar content in peach fruit by a genetic approach: deepening the knowledge of the physiological and metabolic processes regulating sugar accumulation with the aid of new 'omics' tools; developing innovative phenotyping methods to address the strong environmental and genotype × environment effects, even enhancing the accuracy and simplifying the application of ecophysiological models; exploration of the phenotypic variability associated with sugar-related traits, for example, within the Far-East germplasm, to increase the genetic variability and the resolution power of association studies; improvement of the efficiency of *in vitro* transformation for the functional validation of candidate genes; defining more specific targets to be achieved for sugar levels based on fruit type, consumer preferences and perceptions.

ACKNOWLEDGMENTS

This work was supported by MASPES (Italian project aimed at apricot and peach breeding) and FRUITBREEDOMICS (grant 265582 – EC-GAUE, 7th Framework Program: the views expressed in this work are the sole responsibility of the authors and do not necessarily reflect the views of the European Commission).

REFERENCES

1. Cantín CM, Gogorcena Y, Moreno MA. Phenotypic diversity and relationship of fruit quality traits in peach and nectarine [*Prunus persica* (L.) Batsch] breeding progenies. *Euphytica*2010; **171**: 211–226.

2. Crisosto CH, Costa G. Preharvest factors affecting peach quality. In: Layne D R, Bassi D, editors. *The Peach: Botany Production and Uses.* Cambridge: CAB International; 2008. pp536–549.

3. Iglesias I, Echeverría G. Differential effect of cultivar and harvest date on nectarine color, quality and consumer acceptance. *Sci Hortic* 2009; **120**: 41–50.

4. Byrne DH. Trends in stone fruit cultivar development. *Hort Technol* 2005; **15**: 494–500.

5. Crisosto CH. How do we increase peach consumption? *Acta Hort* 2002; **592**: 601–605.

6. Liverani A, Brandi F, Quacquarelli I, Sirri S, Giovannini D. Superior taste and keeping quality are steady goals of the peach breeding activity. *Acta Hort* 2015; **1084**: 179–186.

7. Byrne DH. Peach breeding trends: a worldwide perspective. *Acta Hort* 2002; **592**: 49–59.

8. Callahan AM. Breeding for fruit quality. *Acta Hort* 2003; **622**: 295–302.

9. Reig G, Iglesias I, Gatius F, Alegre S. Antioxidant capacity, quality, and anthocyanin and nutrient contents of several peach cultivars [*Prunus*

persica (L.) Batsch] grown in Spain. *J Agric Food Chem* 2013; **61**: 6344–6357.

10. Byrne DH. Trends in fruit breeding. In: Badenes M L, Byrne DH, editors. *Fruit Breeding. Handbook of Plant Breeding*. New York: Springer; 2012. pp3–36.

11. Iezzoni A, Weebadde C, Luby J *et al.* RosBREED: enabling marker-assisted breeding in Rosaceae. *Acta Hort* 2010; **859**: 389–394.

12. Laurens F, Aranzana MJ, Arús P *et al.* Review of fruit genetics and breeding programmes and a new European initiative to increase fruit breeding efficiency. *Acta Hort* 2012; **92**: 95–102.

13. Byrne DH, Nikolic AN, Burns EE. Variability in sugars, acids, firmness, and colour characteristics of 12 peach genotypes. *J Am Soc Hort Sci* 1991; **116**: 1004–1006.

14. Cantín CM, Gogorcena Y, Moreno MA. Analysis of phenotypic variation of sugar profile in different peach and nectarine [*Prunus persica* (L.) Batsch] breeding progenies. *J Sci Food Agric* 2009; **89**: 1909–1917.

15. Font i Forcada C, Oraguzie N, Igartua E, Moreno MA, Gogorcena Y. Population structure and marker-trait associations for pomological traits in peach and nectarine cultivars. *Tree Genet Genomes* 2013; **9**: 331–349.

16. Lopresti J, Goodwin I, McGlasson B, Holford P, Golding J. Variability in size and soluble solids concentration in peaches and nectarines. In: Janick J, editor. *Horticultural Reviews*. Vol. 42, 1st ed. New Jersey: Wiley-Blackwell; 2014. pp253–311.

17. Predieri S, Ragazzini P, Rondelli R. Sensory evaluation and peach fruit quality. *Acta Hort*2006; **713**: 429–434.

18. Orazem P, Stampar F, Hudina M. Quality analysis of Redhaven peach fruit grafted on 11 rootstocks of different genetic origin in a replant soil. *Food Chem* 2011; **124**: 1691–1698.

19. Delgado C, Crisosto GM, Heymann H, Crisosto CH. Determining the primary drivers of liking to predict consumers' acceptance of fresh nectarines and peaches. *J Food Sci* 2013; **78**: 605–614.

20. Kroger M, Meister K, Kava R. Low-calorie sweeteners and other sugar substitutes: a review of the safety issues. *Compr Rev Food Sci Food Saf* 2006; **5**: 35–47.

21. Robertson JA, Meredith FI. Characteristics of fruit from high- and low-quality peach cultivars.*HortScience* 1988; **23**: 1032–1034.

22. Bassi D, Selli R. Evaluation of fruit quality in peach and apricot. *Adv Hort Sci* 1990; **4**: 107–112.

23. Esti M Messia MC, Sinesio F *et al.* Quality evaluation of peaches and nectarine by electrochemical and multivariate analyses: relationships between analytical measurements and sensory attributes. *Food Chem* 1997; **60**: 659–666.

24. Colaric M, Veberic R, Stampar F, Hudina M. Evaluation of peach and nectarine fruit quality and correlations between sensory and chemical attributes. *J Sci Food Agric* 2005; **85**: 2611–2616.

25. Ortiz A, Lara I, Graell J, Lopez ML, Echeverria G. Sensory acceptance of CA-stored peach fruit. Relationship to instrumental quality parameters. *Acta Hort* 2008; **796**: 225–230.

26. Lopez G, Behboudian MH, Echeverria G, Mata M, Girona J, Marsal J. Instrumental and sensory evaluation of fruit quality for 'Ryan's Sun' peach grown under deficit irrigation.*Horttechnology* 2011; **21**: 712–719.

27. Karakurt Y, Huber DJ, Sherman WB. Quality characteristics of melting and non-melting flesh peach genotypes. *J Sci Food Agric* 2000; **80**: 1843–1853.

28. Crisosto CH, Crisosto GM. Relationship between ripe soluble solids concentration (RSSC) and consumer acceptance of high and low acid melting flesh peach and nectarine (*Prunus persica* (L.) Batsch cultivars. *Postharvest Biol Technol* 2005; **38**: 239–246.

29. Crisosto CH, Crisosto GM, Echeverria G, Puy J. Segregation of peach and nectarine (*Prunus persica* (L.) Batsch cultivars according to their organoleptic characteristics. *Post-harvest Biol Technol* 2006; **39**: 10–18.

30. Kader AA. Fruit maturity, ripening, and quality relationships. *Perishable Handling Newslett*1994; **80**: 2.

31. Hilaire C. The peach industry in France: state of art, research and development. In: Marra F,Sottile F, editors. *Proceedings of First Mediterranean Peach Symposium*. 2003, pp27–34.

32. Testoni A. Momento di raccolta, qualitá, condizionamento e confezionamento delle pesche. In: *Proceedings of the Symposium on 'La peschicoltura Veronese alle soglie del 2000'*. 1995. pp327–354. Italian.

33. Crisosto CH, Crisosto G, Bowerman E. Understanding consumer acceptance of peach, nectarine and plum cultivars. *Acta Hort* 2003; **604**: 115–119.

34. Crisosto CH, Crisosto G, Neri F. Understanding tree fruit quality based on consumer acceptance. *Acta Hort* 2006; **712**: 183–189.

35. Hilaire C, Mathieu V. Le Point sur la qualite gustative des peches et nectarines. *Infos-Ctifl*2004; **201**: 27–31. French.

36. Moriguchi T, Ishizawa Y, Sanada T. Differences in sugar composition in *Prunus persica* fruit and classification by the principal component analysis. *J Jpn Soc Hortic Sci* 1990; **59**: 307–312.

37. Brooks SJ, Moore JN, Murphy JB. Quantitative and qualitative changes in sugar content of peach genotypes [*Prunus persica* (L.) Batsch]. *J Am Soc Hortic Sci* 1993; **118**: 97–100.

38. Lombardo VA, Osorio S, Borsani J *et al.* Metabolic profiling during peach fruit development and ripening reveals the metabolic networks that underpin each developmental stage. *Plant Physiol* 2011; **157**: 1696–1710.

39. Lo Bianco R, Rieger M. Partitioning of sorbitol and sucrose catabolism within peach fruit. *J Am Soc Hortic Sci* 2002; **127**: 115–121.

40. Borsani J, Budde C, Porrini L *et al.* Carbon metabolism of peach fruit after harvest: changes in enzymes involved in organic acid and sugar level modifications. *J Exp Bot* 2009; **60**: 1823–1860.

41. Vizzotto G, Pinton R, Varanini Z, Costa G. Sucrose accumulation in developing peach fruit.*Physiol Plant* 1996; **96**: 225–230.

42. Zanon L, Falchi R, Santi R, Vizzotto G. Sucrose transport and phloem unloading in peach fruit: potential role of two transporters localized in different cell types. *Physiol Plant* 2014;**154**: 179–193.

43. Moing AF, Svanella L, Rolin D, Monet R, Gaudillere JP, Gaudillere M. Compositional changes during fruit development of two peach cultivars differing in juice acidity. *J Am Soc Hortic Sci* 1998; **123**: 770–775.

44. Desnoues E, Gibon Y, Baldazzi V, Signoret V, Genard M, Quilot-Turion B. Profiling sugar metabolism during fruit development in a peach progeny with different fructose-to-glucose ratios. *BMC Plant Biol* 2014; **14**: 336.

45. Chapman GW, Horvat RJ. Changes in non-volatile acids, sugars, pectin and sugar composition of pectin during peach (cv. Monroe) maturation. *J Agric Food Chem* 1990;**8**:383–387.

46. Wu BH, Quilot B, Genard M *et al.* Application of a SUGAR model to analyse sugar accumulation in peach cultivars that differ in glucose-fructose ratio. *J Agric Sci* 2012; **150**: 53–63.

47. Monti L, Bustamante CA, Osorio S *et al.* Metabolic profiling of a range of peach fruit varieties reveals high metabolic diversity and commonalities and differences during ripening.*Food Chem* 2015; **190**: 879–888.

48. Padilla IMG, Golis A, Gentile A, Damiano C, Scorza R. Evaluation of transformation in peach*Prunus persica* explants using green fluorescent protein (GFP) and beta-glucuronidase (GUS) reporter genes. *Plant Cell Tiss Org* 2006; **84**: 309–314.

49. Moing A, Carbonne F, Zipperlin B, Svanella L, Gaudillére JP. Phloem loading in peach: symplastic or apoplastic? *Physiol Plant* 1997; **101**: 489–496.

50. Escobar-Gutierrez AJ, Zipperlin B, Carbonne F, Moing A Gaudillere JP. Photosynthesis, carbon partitioning and metabolite content during drought stress in peach seedlings. *Aust J Plant Physiol* 1998; **25**: 197–205.

51. Zhang LY, Peng YB, Pelleschi-Travier S *et al.* Evidence for apoplasmic phloem unloading in developing apple fruit. *Plant Physiol* 2004; **135**: 574–586.

52. Zhang HP, Wu JY, Tao ST *et al.* Evidence for apoplasmic phloem unloading in pear fruit.*Plant Mol Biol Rep* 2014; **32**: 931–939.

53. Lo Bianco R, Rieger M, Sung SS. Carbohydrate metabolism of vegetative and reproductive sinks in the late-maturing peach cultivar 'Encore'. *Tree Physiol* 1999; **19**: 103–109.

54. Nie PX, Wang XY, Hu LP *et al.* The predominance of the apoplasmic phloem-unloading pathway is interrupted by a symplasmic pathway during Chinese jujube fruit development.*Plant Cell Physiol* 2010; **51**: 1007–1018.

55. Zhang XY, Wang XL, Wang XF *et al.* A shift of phloem unloading from symplasmic to apoplasmic pathway is involved in developmental onset of ripening in grape berry. *Plant Physiol* 2006; **142**: 220–232.

56. Masia A, Zanchin A, Rascio N, Ramina A. Some biochemical and ultrastructural aspects of peach fruit development. *J Am Soc Hortic Sci* 1992; **117**: 808–815.

57. Gao Z, Maurousset L, Lemoine R, Yoo SD, van Nocker S, Loescher W. Cloning, expression, and characterization of sorbitol transporters from developing sour cherry fruit and leaf sink tissues. *Plant Physiol* 2003; **131**: 1566–1575.

58. Verde I, Abbott AG, Scalabrin S *et al.* The high-quality draft genome of peach (*Prunus persica*) identifies unique patterns of genetic diversity, domestication and genome evolution.*Nat Genet* 2003; **45**: 487–494.

59. Wei X, Liu F, Chen C, Ma F, Li M The *Malus domestica* sugar transporter gene family: identifications based on genome and expression profiling related to the accumulation of fruit sugars. *Front Plant Sci* 2014; **5**: 569.

60. Lecourieux F, Kappel C, Lecourieux D *et al.* An update on sugar transport and signalling in grapevine. *J Exp Bot* 2014; **65**: 821–832.

61. Chen LQ, Qu XQ, Hou BH *et al.* Sucrose efflux mediated by SWEET proteins as a key step for phloem transport. *Science* 2012; **335**: 207–211.

62. Chardon F, Bedu M, Calenge F *et al.* Leaf fructose content is controlled by the vacuolar transporter SWEET17 in Arabidopsis. *Current Biol* 2013; **23**: 697–702.

63. Jiang F, Wang Y, Sun H, Yang L, Zhang J, Ma L. Intracellular compartmentation and membrane permeability to sugars and acids at different growth stages of peach. *Sci Hortic*2013; **161**: 210–215.

64. Ruan YL. Sucrose metabolism: gateway to diverse carbon use and sugar signalling. *Annu Rev Plant Biol* 2014; **65**: 33–67.

65. Zhang C, Shen Z, Zhang Y *et al.* Cloning and expression of genes related to the sucrose-metabolizing enzymes and carbohydrate changes in peach. *Acta Physiol Plant* 2013; **35**: 589–602.

66. Moriguchi T, Yamaki S. Purification and characterization of sucrose synthase from peach (Prunus persica) fruit. *Plant Cell Physiol* 1988; **29**: 1361–1366.

67. Roitsch T, Gonzalez MC. Function and regulation of plant invertases: sweet sensations.*Trends Plant Sci* 2004; **9**: 606–613.

68. Tang GQ, Lüscher M, Sturm A. Antisense repression of vacuolar and cell wall invertase in transgenic carrot alters early plant development and sucrose partitioning. *Plant Cell* 1999;**11**: 177–189.

69. Fridman E, Carrari F, Liu YS, Fernie AR, Zamir D. Zooming in on a quantitative trait for tomato yield using interspecific introgressions. *Science* 2004; **305**: 1786–1789.

70. Jin Y, Ni DA, Ruan YL. Posttranslational elevation of cell wall invertase activity by silencing its inhibitor in tomato delays leaf senescence and increases seed weight and fruit hexose level. *Plant Cell* 2009; **21**: 2072–2089.

71. Ohyama A, Ito H, Sato T, Nishimura S, Imai S, Hirai M. Suppression of acid invertase activity by antisense RNA modifies the sugar composition of tomato fruit. *Plant Cell Physiol* 1995;**36**: 369–376.

72. Klann EM, Hall B, Bennett AB. Antisense acid invertase (TIV1) gene alters soluble sugar composition and size in transgenic tomato fruit. *Plant Physiol* 1996; **112**: 1321–1330.

73. Nonis A, Ruperti B, Falchi R, Casatta E, Thamashebi SE, Vizzotto G. Differential expression and regulation of a neutral invertase encoding gene from peach (*Prunus persica*): evidence for a role in fruit development. *Physiol Plant* 2007; **129**: 436–446.

74. Li M, Feng F, Cheng L. Expression patterns of genes involved in sugar metabolism and accumulation during apple fruit development. *PLoS One* 2012; **7**: e33055.

75. Moriguchi T, Sanada T, Yamaki S. Seasonal fluctuations of some enzymes relating to sucrose and sorbitol metabolism in peach fruit. *J Am Soc Hortic Sci* 1990; **115**: 278–281.

76. Genard M, Souty M. Modeling the peach sugar contents in relation to fruit growth. *J Am Soc Hortic Sci* 1996; **121**: 1122–1131.

77. Yamada BK, Niwa N, Shirate K, Yamaki S. cDNA cloning of NAD-dependent sorbitol dehydrogenase from peach fruit and its expression during fruit development. *J Hortic Sci Biotechnol* 2001; **76**: 581–587.

78. Kanayama Y, Kogawa M, Yamaguchi M, Kanahama K. Fructose content and the activity of fructose-related enzymes in the fruit of eating-quality peach cultivars and native-type peach cultivars. *J Jpn Soc Hortic Sci* 2005; **74**: 431–436.

79. Winter H, Huber SC. Regulation of sucrose metabolism higher plants: localization and regulation of activity of key enzymes. *Crit Rev Plant Sci* 2000; **19**: 31–67.

80. Rolland F, Baena-Gonzalez E, Sheen J. Sugar sensing and signaling in plants: conserved and novel mechanisms. *Annu Rev Plant Biol* 2006; **57**: 675–709.

81. Nguyen-Quoc B, Foyer CH. A role for 'futile cycles' involving invertase and sucrose synthase in sucrose metabolism of tomato fruit. *J Exp Bot* 2001; **52**: 881–889.

82. Foulongne M, Pascal T, Arus P, Kervella J. The potential of *Prunus davidiana* for introgression into peach [*Prunus persica* (L.) Batsch] assessed by comparative mapping.*Theoret Appl Genet* 2003; **107**: 227–238.

83. Badenes ML, Cambra M, López MM *et al.* A peach germplasm collection for increasing the genetic diversity in European breeding programs. *Acta Hort* 2015; **1084**: 125–129.

84. Moing A, Poessel JL, Svanella-Dumas L, Loonis M, Kervella J. Biochemical basis of low fruit quality of *Prunus davidiana*, a pest and

disease resistance donor for peach breeding. *J Am Soc Hortic Sci* 2003; **128**: 55–62.

85. Wu B, Quilot B, Kervella J, Genard M, Li S. Analysis of genotypic variation of sugar and acid contents in peaches and nectarines through the Principal Component Analysis. *Euphytica*2003; **132**: 375–384.

86. Quilot B, Wu BH, Kervella J, Genard M, Foulongne M, Moreau K. QTL analysis of quality traits in an advanced backcross between *Prunus persica* cultivars and the wild relative species P. davidiana. *Theor Appl Genet* 2004; **109**: 884–897.

87. Cao K, Wang L, Zhu G, Fang W, Chen C, Luo J. Genetic diversity, linkage disequilibrium, and association mapping analyses of peach (*Prunus persica*) landraces in China. *Tree Genet Genom* 2012; **8**: 975–990.

88. Li XW, Meng XQ, Jia HJ *et al.* Peach genetic resources: diversity, population structure and linkage disequilibrium. *BMC Genet* 2013; **14**: 84.

89. Cao K, Zheng Z, Wang L *et al.* Comparative population genomics reveals the domestication history of the peach, *Prunus persica*, and human influences on perennial fruit crops. *Genome Biol* 2014; **15**: 415.

90. Ma R, Byrne DH, Yu M, Du P, Shen Z. Inbreeding and co-ancestry of the major commercial fresh market peach cultivars in China. *Acta Hort* 2006; **713**: 145–148.

91. Font i Forcada C, Gradziel TM, Gogorcena Y, Moreno MA. Phenotypic diversity among local Spanish and foreign peach and nectarine [*Prunus persica* (L.) Batsch] accessions. *Euphytica*2014; **197**: 261–277.

92. Kwon JH, Jun JH, Nam EY *et al.* Profiling diversity and comparison of Eastern and Western cultivars of *Prunus persica* based on phenotypic traits. *Euphytica* 2015; **206**: 401–415.

93. Dirlewanger E, Moing A, Rothan C *et al.* Mapping QTLs controlling fruit quality in peach [*Prunus persica* (L) Batsch]. *Theor Appl Genet* 1999; **98**: 18–31.

94. Etienne C, Rothan C, Moing A *et al.* Candidate genes and QTLs for sugar and organic acid content in peach [*Prunus persica* (L.) Batsch]. *Theor Appl Genet* 2002; **105**: 145–159.

95. Abidi W, Jiménez S, Moreno MA, Gogorcena Y. Evaluation of antioxidant compounds and total sugar content in a nectarine [*Prunus persica* (L.) Batsch] progeny. *Int J Mol Sci* 2011;**12**: 6919–6935.

96. Souza VAB, Byrne DH, Taylor JF. Heritability, genetic and phenotypic correlations, and predicted selection response of quantitative traits in peach. I. An analysis of several reproductive traits. *J Am Soc Hortic Sci* 1998; **123**: 598–603.

97. Golding JB, Satyan S, Liebenberg C, Walsh K, McGlasson WB. Application of portable NIR for measuring sugar concentrations in peaches. *Acta Hortic* 2006; **713**: 461–464.

98. Genard M, Lescourret F, Gomez L, Habib R. Changes in fruit sugar concentrations in response to assimilate supply, metabolism and dilution: a modeling approach applied to peach fruit (*Prunus persica*). *Tree Physiol* 2003; **23**: 373–385.

99. Crisosto CH, Johnson RS, Luza JG, Crisosto GM. Irrigation regimes affect fruit soluble solids concentration and rate of water loss of 'O'Henry' peaches. *HortScience* 1994; **29**: 1169–1171.

100. Alcobendas R, Mirás-Avalos JM, Alarcón JJ, Nicolás E. Effects of irrigation and fruit position on size, colour, firmness and sugar contents of fruits in a mid-late maturing peach cultivar.*Sci Hortic* 2013; **164**: 340–347.

101. Mimoun MB, Genard M, Besset J. Assimilate allocation to vegetative and fruit growth. *Acta Hort* 1998; **465**: 409–414.

102. Jia HJ, Hirano K, Okamoto G. Effects of fertilizer levels on tree growth and fruit quality of 'Hakuho' peaches (*Prunus persica*). *J Jpn Soc Hort Sci* 1999; **68**: 487–493.

103. Albás ES, Jiménez S, Aparicio J, Betrán JA, Moreno MA. Effect of several peach × almond hybrids rootstocks on fruit quality of peaches. *Acta Hort* 2004; **658**: 321–326.

104. Remorini D, Tavarini S, Degl'Innocenti E, Loreti F, Massai R, Guidi L. Effect of rootstocks and harvesting time on the nutritional quality of peel and flesh of peach fruits. *Food Chem*2008; **110**: 361–367.

105. Orazem P, Stampar F, Hudina M. Fruit quality of Redhaven and Royal Glory peach cultivars on seven different rootstocks. *J Agric Food Chem* 2011; **59**: 9394–9401.

106. Farina V, Lo Bianco R, Inglese P. Vertical distribution of crop load and fruit quality within vase and Y-shaped canopies of 'Elegant Lady' peach. *HortScience* 2005; **40**: 587–591.

107. Luchsinger L, Ortin P, Reginato G, Infante R. Influence of canopy position on the maturity and quality of 'Angelus' peaches. *Acta Hort* 2002; **592**: 515–521.

108. Kumar M, Rawat V, Rawat JMS, Tomar YK. Effect of pruning intensity on peach yield and fruit quality. *Sci Hortic* 2010; **125**: 218–221.

109. Marini RP, Trout JR. Sampling procedures for minimizing variation in peach fruit quality. *J Am Soc Hort Sci* 1984; **109**: 361–364.

110. Genard M, Bruchou C. Multivariate analysis of within-tree factors accounting for the variation of peach fruit quality. *Sci Hortic* 1992; **52**: 37–51.

111. Grossman YL, DeJong TM. Maximum fruit growth potential and seasonal patterns of resource dynamics during peach growth. *Ann Bot* 1995; **75**: 553–560.

112. Grossman YL, DeJong TM. Maximum fruit growth potential and seasonal patterns of resources dynamics during peach growth. *Ann Bot* 1995; **75**: 561–567.

113. Golding JB, Pristijono P, Spohr L, McGlasson B. Variability in peach and nectarine eating quality. *Acta Hortic* 2012; **962**: 499–507.

114. Forlani M, Basile B, Cirillo C, Iannini C. Effects of harvest date and fruit position along the tree canopy on peach fruit quality. *Acta Hortic* 2002; **592**: 459–466.

115. Culpepper CW, Caldwell JS. The canning quality of certain commercially important eastern peaches. *US Dept Agric Tech Bull* 1930; **196**.

116. Morandi B, Corelli-Grappadelli L, Rieger M, Lo Bianco R. Carbohydrate availability affects growth and metabolism in peach fruit. *Physiol Plant* 2008; **133**: 229–241.

117. Bertin N, Bussieres P, Genard M. Ecophysiological models of fruit quality: a challenge for peach and tomato. *Acta Hortic* 2006; **718**: 633–645.

118. Boote KJ, Kropff MJ, Bindraban PS. Physiology and modelling of traits in crop plants: Implications for genetic improvement. *Agric Syst* 2001; **70**: 395–420.

119. Génard M, Bertin N, Gautier H, Lescourret F, Quilot B. Virtual profiling: a new way to analyse phenotypes. *Plant J* 2010; **62**: 344–355.

120. Quilot B, Genard M, Lescourret F, Kervella J. Simulating genotypic variation of fruit quality in an advanced peach × *Prunus davidiana* cross. *J Exp Bot* 2005; **56**: 3071–3081.

121. Grechi I, Hilgert N, Genard M, Lescourret F. Assessing the peach fruit refractometric index at harvest with a simple model based on fruit growth. *J Am Soc Hortic Sci* 2008; **133**: 178–187.

122. Quilot B, Genard M, Kervella J, Lescourret F. Analysis of genotypic variation in fruit flesh total sugar content via an ecophysiological model applied to peach. *Theor Appl Genet* 2004;**109**: 440–449.

123. Lescourret F, Genard M. A virtual peach fruit model simulating changes in fruit quality during the final stage of fruit growth. *Tree Physiol* 2005; **25**: 1303–1315.

124. Bertin N, Martre P, Genard M, Quilot B, Salon C. Under what circumstances can process-based simulation models link genotype to phenotype for complex traits? Case-study of fruit and grain quality traits. *J Exp Bot* 2010; **61**: 955–967.

125. Quilot B, Kervella J, Genard M, Lescourret F. Analysing the genetic control of peach fruit quality through a ecophysiological model combined with a QTL approach. *J Exp Bot* 2005;**56**: 3083–3092.

126. Struik PC, Yin X, de Visser P. Complex quality traits: now time to model. *Trends Plant Sci*2005; **10**: 513–516.

127. Zorrilla-Fontanesi Y, Cabeza A, Domínguez P *et al.* Quantitative trait loci and underlying candidate genes controlling agronomical and fruit quality traits in octoploid strawberry (Fragaria × ananassa). *Theor Appl Genet* 2011; **123**: 755–778.

128. Guan Y, Peace C, Rudell D, Verma S, Evans K. QTLs detected for individual sugars and soluble solids content in apple. *Mol Breeding* 2015; **35**: 1–13.

129. Wang D, Karle R, Iezzoni AF. QTL analysis of flower and fruit traits in sour cherry. *Theor Appl Genet* 2000; **100**: 535–544.

130. Socquet-Juglard D, Christen D, Devénes G, Gessler C, Duffy B, Patocchi A. Mapping architectural, phenological, and fruit quality QTLs in apricot. *Plant Mol Biol Rep* 2013; **31**: 387–397.

131. Balibrea ME, Martinez-Andujar C, Cuartero J, Bolarin MC Perez-Alfocea F. The high fruit soluble sugar content in wild Lycopersicon species and their hybrids with cultivars depends on sucrose import during ripening rather than on sucrose metabolism. *Funct Plant Biol* 2006;**33**: 279–288.

132. Petreikov M, Yeselson L, Shen S *et al.* Carbohydrate balance and accumulation during development of near-isogenic tomato lines differing in the AGPase-L1 allele. *J Am Soc Hortic Sci* 2009; **134**: 134–140.

133. Zanor MI, Osorio S, Nunes-Nesi A *et al.* RNA interference of LIN5 in tomato confirms its role in controlling Brix content, uncovers the influence of sugars on the level of fruit hormones, and demonstrates the importance of sucrose cleavage for normal fruit development and fertility. *Plant Physiol* 2009; **150**: 1204–1218.

134. Salazar JA, Ruiz D, Campoy JA *et al.* Quantitative trait loci (QTL) and Mendelian trait loci (MTL) analysis in *Prunus*: a breeding perspective and beyond. *Plant Mol Biol Rep* 2013; **32**: 1-18.

135. Abbott AG, Rajapakse S, Sosinski B *et al.* Construction of saturated linkage maps of peach crosses segregating for characters controlling fruit quality, tree architecture and pest resistance. *Acta Hortic* 1998; **465**: 41–50.

136. Dirlewanger E, Cosson P, Renaud C, Monet R, Poessel JL, Moing A. New detection of QTLs controlling major fruit quality components in peach. *Acta Hortic* 2005; **713**: 65–72.

137. Eduardo I, Pacheco I, Chietera G *et al.* QTL analysis of fruit quality traits in two peach intraspecific populations and importance of maturity date pleiotropic effect. *Tree Genet Genomes* 2011; **7**: 323–335.

138. Verde I, Quarta R, Cedrola C, Dettori MT. QTL analysis of agronomic traits in a BC1 peach population. *Acta Hortic* 2002; **592**: 291–297.

139. Fresnedo-Ramírez J, Bink M, de Weg E *et al.* QTL mapping of pomological traits in peach and related species breeding germplasm. *Mol Breeding* 2015; **35**: 166.

140. Illa E, Eduardo I, Audergon J *et al.* Saturating the *Prunus* (stone fruits) genome with candidate genes for fruit quality. *Mol Breeding* 2011; **28**: 667–682.

141. Pirona R, Eduardo I, Pacheco I *et al.* Fine mapping and identification of a candidate gene for a major locus controlling maturity date in peach. *BMC Plant Biol* 2013; **13**: 166.

142. Souza VAB, Byrne DH, Taylor JF. Predicted breeding values for nine plant and fruit characteristics of 28 peach genotypes. *J Am Soc Hortic Sci* 2000; **125**: 460–465.

143. Prudent M, Causse M, Genard M, Tripodi P, Grandillo S, Bertin N. Genetic and physiological analysis of tomato fruit weight and composition: influence of carbon availability on QTL detection. *J Exp Bot* 2009; **60**: 923–937.

144. DeJong TM. Developmental and environmental control of dry-matter partitioning in peach.*HortScience* 1999; **34**: 1037–1040.

145. Lopresti J, Holford P, McGlasson B, Goodwin I, Golding J. Studies on the relationship between sugar concentration and cell number and size distribution in mesocarp of nectarine fruit. *Acta Hortic* 2015; **1084**: 667–674.

146. Génard M, Reic M, Lobit P, Besset J. Correlations between sugar and acid content and peach growth. *J Hortic Sci Biotechnol* 1999; **74**: 772–776.

147. Kano Y. Effects of mechanically restricting melon (*Cucumis melo* L.) fruit growth on cell size and sugar accumulation. *Environ Control Biol* 2009; **47**: 13–19.

148. Morandi B, Rieger M, Corelli-Grappadelli L. Vascular flows and transpiration affect peach [*Prunus persica* (L.) Batsch.] fruit daily growth. *J Exp Biol* 2007; **58**: 3941–3947.

149. Kobashi K, Gemma H, Iwahori S. Abscisic acid content and sugar metabolism of peaches grown under water stress. *J Am Sci Hortic Sci* 2000; **125**: 425–428.

150. DeJong TM, Grossman YL. Quantifying sink and source limitations on dry matter partitioning to fruit growth in peach trees. *Physiol Plant* 1995; **95**: 437–443.

151. Wen IC, Koch KE, Sherman WB. Comparing fruit and tree characteristics of two peaches and their nectarine mutants. *J Am Soc Hortic Sci* 1995; **120**: 101–106.

152. Wen IC, Sherman WB, Koch KE. Heritable pleiotropic effects of the nectarine mutant to peach. *J Am Soc Hortic Sci* 1995; **120**: 721–725.

153. Dagar A, Pons Puig C, Marti Ibanez C *et al.* Comparative transcript profiling of a peach and its nectarine mutant at harvest reveals differences in gene expression related to storability.*Tree Genet Genomes* 2012; **9**: 223–235.

154. Vendramin E, Pea G, Dondini L *et al.* A unique mutation in a MYB gene co-segregates with the nectarine phenotype in peach. *PLoS One* 2014; **9**: e90574.

155. Osorio S, Ruan YL, Fernie AR. An update on source-to-sink carbon partitioning in tomato.*Front Plant Sci* 2014; **5**: 516.

156. Ma R, Yu M, Du P, Guo H, Song H. Evaluation of germplasm resources and breeding of flat peach. *Acta Hortic* 2003; **620**: 161–167.

157. Wang L, Zhu G, Fang W, Cao K, Chen C. Comparison of heritable pleiotropic effects of the glabrous and flat shape traits of peach. *Can J Plant Sci* 2010; **90**: 367–370.

158. Robertson JA, Horvat JA, Lyon BG, Meredith FI, Senter SD, Okie WR. Comparison of quality characteristics of selected yellow- and white-fleshed peach cultivar. *J Food Sci* 1990; **15**: 323–334.

159. Wu BH, Quilot B, Genard M, Kervella J, Li SH. Changes in sugar and organic acid concentrations during fruit maturation in peaches, *P. davidiana* and hybrids as analyzed by principal component analysis. *Sci Hort* 2005; **103**: 429–439.

160. Bliss FA, Arulsekar S, Foolad MR *et al.* An expanded genetic linkage map of Prunus based on an interspecific cross between almond and peach. *Genome* 2002; **45**: 520–529.

161. Miller MR, Dunham JP, Amores A, Cresko WA, Johnson EA. Rapid and cost-effective polymorphism identification and genotyping using restriction site associated DNA (RAD) markers. *Genome Res* 2007; **17**: 240–248.

162. Elshire RJ, Glaubitz JC, Sun Q *et al.* A robust, simple genotyping-by-sequencing (GBS) approach for high diversity species. *PLoS One* 2011; **6**: e19379.

163. Meuwissen TH, Hayes BJ, Goddard ME. Prediction of total genetic value using genome-wide dense marker maps. *Genetics* 2001; **157**: 1819–1829.

164. Takagi H, Abe A, Yoshida K *et al.* QTL-seq: rapid mapping of quantitative trait loci in rice by whole genome resequencing of DNA from two bulked populations. *Plant J* 2013; **74**: 174–183.

165. Aranzana MJ, Abbassi K, Howad W, Arus P. Genetic variation, population structure and linkage disequilibrium in peach commercial varieties. *BMC Genet* 2010; **11**: 69.

166. Tester M, Langridge P. Breeding technologies to increase crop production in a changing world. *Science* 2010; **327**: 818–822.

Chapter 5

Low soil water content during growth contributes to preservation of green colour and bioactive compounds of cold-stored broccoli (*Brassica oleraceae* L.) florets

Sarah L.P. Cogo, Fábio C. Chaves, Manoel A. Schirmer, Rui C. Zambiazi, Leonardo Nora, Jorge A. Silva, Cesar V. Rombaldi

Universidade Federal de Pelotas, Faculdade de Agronomia Eliseu Maciel, Departamento de Ciência e Tecnologia Agroindustrial, Caixa Postal 354, CEP 90010-900, Pelotas, RS, Brazil

ABSTRACT

Broccoli, cultivated under low (0.40 MPa) and normal (0.04 MPa, equivalent to field capacity) soil water content, and stored under low (1 °C) and room (23 °C) temperature, was assessed for changes in colour, bioactive compounds, and antioxidant activity. Results demonstrated a significant interaction between cultivation and storage conditions. Low soil water content during plant growth and postharvest cold storage were the conditions that, combined, gave the best preservation of colour, antioxidant activity, and L-ascorbic acid and 5-methyl-tetrahydrofolate contents. Carotenoid preservation was dependent on postharvest storage conditions while the contents of phenolic compounds were reduced over time, independent of cultivation and postharvest storage conditions.

KEYWORDS

Abiotic stress; Secondary metabolite; Antioxidant activity

1. INTRODUCTION

Broccoli is an important source of L-ascorbic acid, folates, glucosinolates, flavonoids, vitamins and dietary fibres (Vallejo et al., 2003). However, significant postharvest loss of nutritional and functional properties, including that of bioactive compounds such as ascorbic acid, glucosinolates, and folic acid, have been shown to occur (Jones et al., 2006 and Moreno et al., 2006). Attempts to minimize the problem include the use of cold storage (Able et al., 2005), heat treatment (Funamoto et al., 2003), improvement of cytokinin synthesis through transgenic plants (Chen et al., 2001), application of cytokinin (Costa et al., 2005), ethanol vapours (Asoda et al., 2009), 1-MCP treatment (Able et al., 2002), storage under modified (Rai et al., 2008) or controlled atmosphere (Watada et al., 1996 and Eason et al., 2007), and UV-C treatment (Costa et al., 2006). Among these methods, cold storage with or without controlled or modified atmosphere is the most widely employed strategy (Toivonen, 1997).

Recently we demonstrated that low soil water content (0.40 MPa of soil water tension) during broccoli growth led to leaf size reduction, without affecting weight or yield, and contributed to the maintenance of green colour, possibly due to induced cytokinin synthesis (Zaicovski et al., 2008). Cytokinins, which are considered senescence inhibitors acting as ethylene antagonists and protectors of membranes, mitochondria, and plastid metabolism are also known to be induced in response to stresses (Xu and Huang, 2009). However, Wurr et al. (2002) observed that severe stress (−0.6 MPa of soil water pressure) led to negative effects on broccoli yield and morphology.

Cultivation under biotic and abiotic stress conditions can stimulate synthesis of bioactive compounds (Oh et al., 2009), mostly associated with antioxidant systems linked to plant defense mechanisms (Mittler, 2002).

Fortier et al. (2010) observed that broccoli grown under nitrogen and water stress had an increased content of phenolic compounds. In addition, contents of bioactive compounds in fruit and vegetables can also be influenced by genotype, maturity at harvest, weather, cultivation, and storage conditions (Podsedek, 2007).

In this context, we have investigated the influence of soil water content during plant growth and postharvest storage conditions on colour, antioxidant activity, and contents of bioactive compounds of broccoli florets, including carotenoids, phenolic compounds, L-ascorbic acid, and 5-methyl-tetrahydrofolate.

2. MATERIALS AND METHODS

2.1. Plant material

'Green Star' broccoli (*Brassica oleraceae* L. var. italica) was cultivated in a greenhouse, with air temperatures of 15.5–24.9 °C and natural light of 6.7–15.7 mol m^{-2} day^{-1}, in Pelotas, RS, Brazil, from April through September of 2009. The soil was an Acrisol (dark red podzolic by Brazilian soil taxonomy and palendult by US soil taxonomy), with 530 g kg^{-1} of sand, 210 g kg^{-1} of silt and 230 g kg^{-1} of clay, pH 6.2, organic matter 3.8%, phosphorus 29.7 mg dm^{-3}, potassium 180 mg dm^{-3}, sodium 8 mg dm^{-3}, aluminium 1.3 mmol dm^{-3}, magnesium 3 mmol dm^{-3}, and boron 0.06 mg dm^{-3}. Soil fertility was adjusted using 60 kg ha^{-1} of nitrogen (ammonium sulphate), 300 kg ha^{-1} of P_2O_5 (simple superphosphate), 240 kg ha^{-1} of K_2O (potassium chloride) and 6 kg ha^{-1} of boron (borax). In addition, three nitrogen applications (20 kg ha^{-1} of nitrogen as urea) were performed at 15, 30, and 45 d after transplanting. Soil temperature measured at 20 cm varied from 19.8 to 20.9 °C.

Broccoli was cultivated under low (L) and normal (N) soil water contents (SWC), corresponding to soil water tensions of 0.40 ± 0.02 MPa and 0.04 ± 0.02 MPa (equivalent to field capacity), respectively. SWC were established 7 d before transplanting (plants with four or five leaves) and maintained until harvest, using an automated irrigation and monitoring system (Irrigás PRO® and MDI-10S®), as described by Zaicovski et al. (2008). One dripper for every two plants, and SWC sensors installed at 20 cm bellow soil surface, were used. The cultivation system described above is commonly used for commercial production of high quality broccoli in Brazil.

65 d after transplanting, 140 broccoli heads from each SWC were selected for uniformity, harvested at sunset (between 6 and 7 pm), weighed (450–500 g), and placed in ventilated plastic boxes (14 broccoli heads per box). Out of the 140 broccoli heads from each SWC (L and N), half was stored at room temperature (RT), at 23 ± 2 °C and 75 ± 5% relative humidity (RH), (LRT and NRT treatments), and the other half was kept under cold storage (CS), at 1 ± 1 °C and 85 ± 5% RH, (LCS and NCS treatments). Daily, for 7 d, 9 broccoli heads were sampled randomly from each treatment (LCS, LRT, NCS and NRT), pooled in triads, weighed, measured for colour, cut up into 2 cm long florets, mixed well, frozen in liquid nitrogen and stored at −80 °C until analysis.

2.2. Analyses

2.2.1. Colour

Colour of intact broccoli heads was measured with a chromameter (Minolta™ CR-300) and reported as hue angle [$h = \tan^{-1}(b/a)$ when $a > 0$ and $b > 0$ or $h = 180 + \tan^{-1}(b/a)$ when $a < 0$ and $b > 0$)] as previously described by Zaicovski et al. (2008).

2.2.2. Carotenoids

Broccoli (5 g fresh fw) was finely ground in liquid nitrogen, suspended in 10 mL of 80% acetone (v/v), stirred for 15 min and filtered (the extraction was repeated three times). The filtrate was centrifuged at 10,000 × *g* for 10 min and the supernatant brought to 40 mL with the acetone solution. Absorbance was measured at 647, 663 and 470 nm in an UV/Vis spectrophotometer. Carotenoids content was determined using the equations described by Lichtenthaler (1987) and expressed in μg g^{-1} fw.

2.2.3. L-Ascorbic acid

Broccoli (5 g fw) was finely ground in liquid nitrogen, suspended in 30 mL of a cold (4 °C) metaphosphoric acid solution (4.5%, w/v in water), and stored at 4 °C for 1 h in the dark and then brought to 50 mL with deionized water.

The sample was filtered and the filtrate centrifuged at 12,000 × *g* for 10 min at 4 °C. The supernatant was filtered through a 0.45 μm Durapore membrane and 25 μL was injected in a HPLC Shimadzu system, using a reverse phase Shim-Pak CLC-ODS column (4.6 mm × 150 mm × 5 μm). An elution gradient started at 100% acetic acid (0.1%, v/v, A), then linearly reduced at 5 min to 98% of A and 2% of methanol (B), then held for 2 min and returned to initial conditions at 10 min. Flow rate was 0.8 mL min^{-1} and the UV detector was set at 254 nm. Identification was based on retention time comparison to an L-ascorbic acid standard. Quantification was based on external standard calibration curve and results were expressed as L-ascorbic acid in mg 100 g^{-1} fw.

2.2.4. 5-Methyl-tetrahydrofolate (5-mthf)

Broccoli (5 g fw) was finely ground in liquid nitrogen, then 15 mL of ammonium acetate solution (0.05 M) was added, and the extract placed in a sonic bath for 10 min. The mixture was transferred to a volumetric flask, to which 500 μL of trichloroacetic acid was added and final volume brought to 10 mL with the ammonium acetate solution. After homogenization, the sample was centrifuged for 20 min at 12,000 × *g* and filtered through a 0.45 μm Durapore membrane. An aliquot of 10 μL was injected in the previously described HPLC system. Analytical conditions were as follows: flow rate of 0.8 mL min^{-1}, initially 100% of acetic acid (2%, v/v, pH 2.8, A), at 20 min 70% A and 30% acetonitrile (B), at 25 min 100% A, held until 40 min. The fluorescence detector excitation wavelength was set at 290 nm and emission at 360 nm. 5-Methyl-tetrathydrofolate (5-mthf) identification was based on retention time comparison with a standard. Quantification was obtained from an external standard calibration curve and results were expressed as μg 100 g^{-1} fw.

2.2.5. Phenolic compounds

Broccoli (5 g fw) finely ground in liquid nitrogen was suspended in 20 mL of cold methanol (4 °C) and stirred for 24 h at 4 °C in the dark; then centrifuged at 4 °C for 15 min at 12,000 × *g*. Phenolic compounds were measured at 760 nm and quantification was expressed as gallic acid equivalent in mg g^{-1} fw, following the Folin–Ciocalteu method.

2.2.6. Antioxidant activity

Antioxidant activity was determined using 2, 2-diphenyl-1-picrylhydrazyl (DPPH), as described by Brand-Williams et al. (1995). Briefly, broccoli (5 g fw) was finely ground in liquid nitrogen, suspended in 20 mL of cold methanol (4 °C), stirred for 24 h at 4 °C in the dark; and centrifuged at 4 °C for 15 min at 12,000 × *g*. The reaction was carried out using 100 μL of the supernatant and 3.9 mL of DPPH (100 μM) in methanol. After 60 min (exponential phase of the reaction curve) absorbance (Abs) was measured at 517 nm and the radical sequestering capacity was expressed as percent inhibition, calculated as follows: %Inhibition = [(Abs blank − Abs sample)/Abs blank] × 100.

2.3. Experimental design and statistical analysis

The experiment was set up in a completely randomized design with three replicates per treatment. Three experimental factors were considered: (1) soil water content (SWC) during plant growth, with two levels (L, 0.40 MPa and N, 0.04 MPa); (2) postharvest storage condition (PSC), with two levels (CS, 1 °C at 85% RH and RT, 23 °C at 75% RH); and (3) postharvest storage period (PSP), with eight levels (0, 1, 2, 3, 4, 5, 6 and 7 d). Antioxidant activity expressed as percentage was normalized according to the equation$f(x)$ = arcsine $X^{1/2}$. Statistical analysis of the data was performed using SAS version 9.1 for Windows (SAS Institute, Cary, NC, USA) and Metrixus (Élin Duxus, São Paulo, SP, Brazil), an *add-in* program for Microsoft Office Excel. Using SAS, data were subjected to an analysis of variance, performed using the F test. Using Metrixus, data were subjected to Pearson's correlation test and principal component analysis (PCA).

3. RESULTS

From 0 d through 3 d, as expected, LCS and NCS broccoli were greener than LRT and NRT broccoli (Fig. 1a), reinforcing that cold storage contributes to a better preservation of the green colour. However, from 4 d through 7 d the response was dependent on the SWC. As mentioned above, no yellowing was observed in LCS broccoli (hue angle > 120) for at least 6 d, while in broccoli stored at room temperature (LRT and NRT) yellowing was observed 3 d after harvest. These results indicate that cold storage potentiates the effect of low SWC during plant growth in the preservation of the green colour. Moreover, a significant interaction between SWC during plant growth and PSC was observed (Table 1). While yield was not affected by SWC, postharvest weight loss was affected by PSC only. On average, cold-stored broccoli lost 0.76% of its weight per day, for a total of 5.31% weight loss, while broccoli stored at room temperature lost 4.75% of its weight per day, for a total of 33.24% weight loss (data not shown).

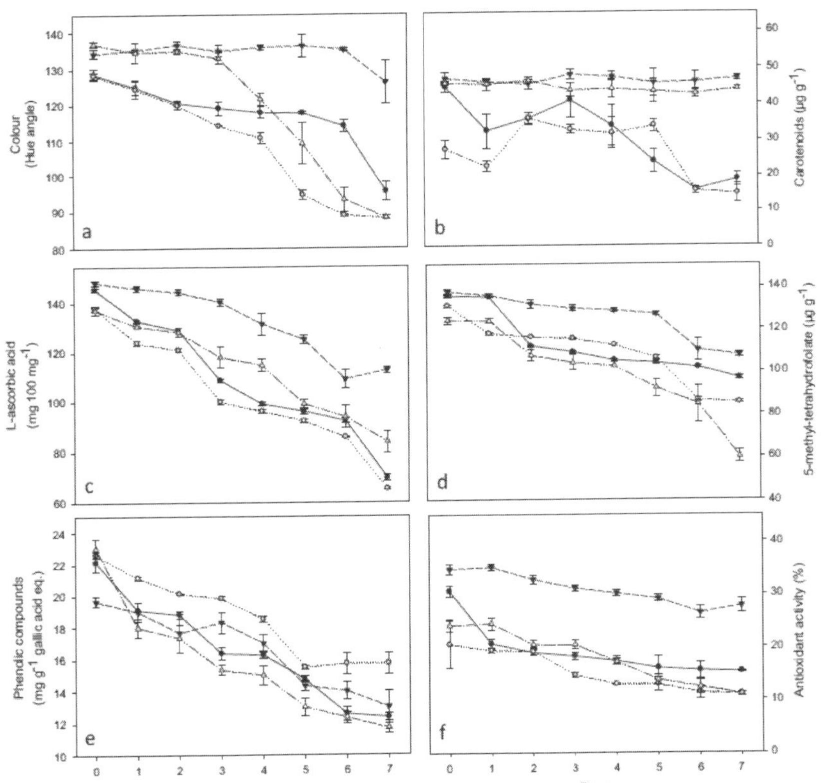

Figure1. Colour (a), carotenoids (b), L-ascorbic acid (c), 5-methyl-tetrahydrofolate (d), phenolic compounds (e), and antioxidant activity (f) of 'Green Star' broccoli cultivated under low (L) and normal (N) soil water content, 0.40 ± 0.02 MPa and 0.04 ± 0.02 MPa (equivalent to field capacity), respectively, and postharvest stored under room temperature (RT), at 23 ± 2 °C and 75 ± 5% relative humidity (● LRT and ○ NRT), and under cold storage (CS), at 1 ± 1 °C and 85 ± 5% relative humidity (▾ LCS and △ NCS), for 7 d. Results (charts b–f) are expressed on fresh weight basis. Error bars represent standard error of the mean (*n* = 3).

Table 1. Interactions between soil water content (SWC), postharvest storage condition (PSC), and postharvest storage period (PSP) for all the measured variables, including colour, carotenoids, phenolic compounds, L-ascorbic acid, 5-methyl-tetrahydrofolate (5-mthf), and antioxidant activity, determined by analysis of variance (ANOVA).

Source	*F* values					
	Colour	Carotenoids	Phenolics compounds	L-Ascorbic acid	5-Mthf	Antioxidant activity
SWC*PSC	17.85***	1.11[NS]	83.21***	80.41***	195.94***	79.67***
SWC*PSP	36.92***	2.66*	1.56[NS]	1.79[NS]	9.72***	2.36*
PSC* PSP	2.59*	10.33***	2.67*	24.13***	4.39**	1.92[NS]
SWC*PSC*PSP	6.47**	2.82*	4.39**	3.44*	11.84***	2.33*

NS: non-significance.
* Indicates significance at the 0.05 level.
** Indicates significance at the 0.001 level.
*** Indicates significance at the 0.0001 level.

PSC had greater impact than SWC on carotenoid preservation (Fig. 1b), without a significant interaction between these experimental factors (Table 1). Broccoli kept under CS (LCS and NCS) maintained higher contents of carotenoids than those stored at RT (LRT and NRT), independent of SWC. During 7 d of storage, the contents of carotenoids were not significantly affected in broccoli under CS (LCS and NCS) while in broccoli under RT (LRT and NRT) an average loss of 84% occurred (Fig. 1b).

CS alone was not enough to avoid significant losses of L-ascorbic acid. Although L-ascorbic acid content dropped during storage, in LCS broccoli the losses were limited (Fig. 1c). This indicates that preservation of this compound depends on the association of SWC and PSC (Table 1). After 7 d of storage, LCS broccoli had lost about 24% of L-ascorbic acid, while NCS broccoli had lost about 40% and LRT and NRT broccoli had lost 52% each. Following a similar trend, 5-mthf preservation was a consequence of the interaction between SWC and PSC (Fig. 1d and Table 1). At the end of 7 d of storage, LCS broccoli (best treatment) and NCS broccoli (worst treatment) had lost about 21% and 52% of 5-mthf, respectively.

Carotenoids, L-ascorbic acid, and 5-mthf contents and antioxidant activity in LCS broccoli were higher than in other treatments (Fig. 1). In contrast to the other measured variables, the content of phenolic compounds at harvest was higher in broccoli grown under normal SWC than in broccoli grown under low SWC (Fig. 1e). During storage, the content of phenolic compounds was reduced in all treatments over time. SWC (L and N) or PSC (RT and CS) induced no clear effect on phenolic content.

Principal component analysis (PCA) of the data, led to a reduction of the variables and establishment of two principal components (PC1 and PC2) that represented the majority of the observed variation (87.12%). In this manner it was possible to identify a general trend of the treatments considering the contribution of all measured variables (Fig. 2). PC1 represented 71.06% of the total variation, with major contribution from the variables colour, carotenoids, and antioxidant activity; while PC2 represented 16.06% of total the variation, with main contribution from phenolic compounds and 5-mthf. Broccoli stored under CS and RT formed groupings (CS solid line and RT dotted line, Fig. 2) separated along the *Y*-axis represented by PC2. LRT broccoli did not separate from NRT broccoli, indicating a limited effect of the SWC on broccoli stored at room temperature. On the other hand, NCS broccoli separated from LCS broccoli, illustrating the influence of the SWC on cold stored broccoli. Within groupings (circled in Fig. 2), a continuous progression from 0 d to 7 d indicated a postharvest storage period influence, which was correlated with the general trend observed for all variables in which values decreased over time (Fig. 1).

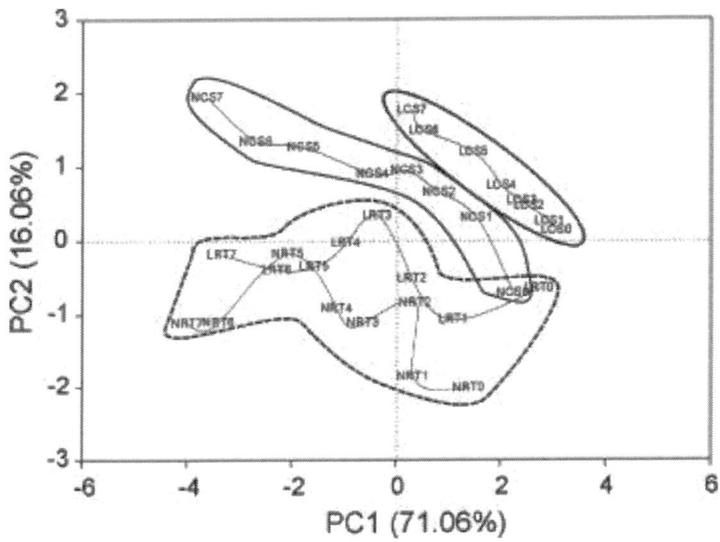

Figure 2. Principal component analysis (PCA) of 'Green Star' broccoli cultivated under low (L) and normal (N) soil water content, 0.40 ± 0.02 MPa and 0.04 ± 0.02 MPa (equivalent to field capacity), respectively, and postharvest stored under room temperature (RT), at 23 ± 2 °C and 75 ± 5% relative humidity (LRT and NRT, 0–7 d), and under cold storage (CS), at 1 ± 1 °C and 85 ± 5% relative humidity (LCS and NCS, 0–7 d).

The highest correlations found for the measured variables were between L-ascorbic acid and colour (0.87), L-ascorbic acid and 5-mthf (0.85), and L-ascorbic acid and antioxidant activity (0.83); meanwhile the lowest correlations were between phenolic compounds and carotenoids (0.08), followed by carotenoids and 5-mthf (0.34), and phenolic compounds and antioxidant activity (0.37) (Table 2).

Table 2. Pearson's correlations between the measured variables: colour, L-ascorbic acid, antioxidant activity, 5-methyl-tetrahydrofolate (5-mthf), carotenoids and phenolic compounds.

Correlations	Colour	L-Ascorbic acid	Antioxidant activity	5-Mthf	Carotenoids	Phenolic compounds
Colour	1.000					
L-Ascorbic acid	0.870	1.000				
Antioxidant activity	0.826	0.835	1.000			
5-Mthf	0.803	0.851	0.786	1.000		
Carotenoids	0.603	0.616	0.633	0.337	1.000	
Phenolic compounds	0.487	0.673	0.375	0.718	0.079	1.000

4. DISCUSSION

One of the most easily perceived alterations in broccoli leading to loss of commercial value is yellowing which occurs 3–4 d after harvest if maintained at room temperature (Zaicovski et al., 2008). Widely documented in the literature (Zaicovski et al., 2008 and Asoda et al., 2009), this problem has various suggested methods of prevention, including CS (Starzynska et al., 2003, Vallejo et al., 2003 and Able et al., 2005), with or without controlled or modified atmosphere (Watada et al., 1996, Rakotonirainy et al., 2001, Serrano et al., 2006, Eason et al., 2007 and Rai et al., 2008), which contributes to prevent the degradation of chlorophylls caused by enzymes such as chlorophyllase (Hortensteiner, 2006), peroxidase (Funamoto et al., 2003), pheophytinase (Schelbert et al., 2009), and pheophorbide oxygenase (Pruzinska et al., 2005).

In the present study, CS prevented yellowing of 'Green Star' broccoli cultivated under low SWC only; demonstrating an interaction between SWC during plant growth and PSC (Table 1). Zaicovski et al. (2008) observed that low SWC during broccoli growth induced formation of new roots and accumulation of cytokinins in the florets, resulting in better preservation of green colour, possibly a consequence of the protective action of this hormone. In the study of Wurr et al. (2002), water stressed broccoli were also greener and more turgid; however, the more severe stress conditions (−0.6 MPa) negatively affected morphology and yield. Cytokinin application (Costa et al., 2005) or cytokinin synthesis induction by an *ipt* transgene (Chen et al., 2001) protects the inflorescences from yellowing. In addition, Synkova et al. (2006) has shown that tobacco over-expressing an *ipt* gene had increased defense responses, possibly leading to a better preservation of cell homeostasis. 'Green Star' broccoli cultivated under low SWC showed a ten-fold increase in trans-zeatin riboside when compared to broccoli cultivated under normal SWC (Zaicovski et al., 2008). Xu and Huang (2009) also observed that creeping bentgrass treated with trans-zeatin riboside became more tolerant to abiotic stress.

Low SWC contributed synergistically with CS in the maintenance of colour (Fig. 1a) and some bioactive molecules, such as L-ascorbic acid (Fig. 1c), 5-mthf (Fig. 1d) and antioxidant activity (Fig. 1f), but not in the maintenance of carotenoids (Fig. 1b) and phenolic compounds (Fig. 1e). CS was a determinant in the conservation of carotenoids (Fig. 1b), but neither CS nor low SWC was effective in preventing losses of phenolic compounds (Fig. 1e).

Carotenoid contents were kept relatively constant during the course of 7 d under CS (Fig. 1b), which is particularly important from a nutritional point of view since carotenoids are an important source of vitamin A precursors (Zhang and Hamauzu, 2004 and Jones et al., 2006). From a physiological perspective, carotenoids protect chloroplasts against photooxidation, preventing chlorophyll degradation and avoiding the formation of free radicals in plastids (Aluru et al., 2009). In this study a large variation in carotenoid contents was observed, although, the exact causes were not identified. Senescence and physical injuries might have accounted for the observed losses.

Contents of phenolic compounds were lower in broccoli cultivated under low SWC, contrasting with the initial hypothesis based on work by Mittler (2002), Oh et al. (2009) and Fortier et al. (2010), who proposed increased accumulation of these compounds upon abiotic stresses. However, not all abiotic stresses have beneficial impacts or always give the same response. For example, under low SWC, accumulation of L-ascorbic acid is increased in some fruit and vegetables but is reduced in strawberry and leafy greens (Cisneros-Zevallos, 2003). Phenolic compounds were dramatically reduced under all treatments, independent of SWC or PSC. Leja et al. (2001) and Zhang and Hamauzu (2004) have reported broccoli with phenolic compounds contents above 34.5 mg 100 g^{-1} fw while in this study, they ranged from 19.7 to 23.0 mg 100 g^{-1} fw at harvest and from 11.7 to 15.7 mg 100 g^{-1} fw after 7 d of storage. Reduction in phenolic compounds during storage could be attributed to the bioconversion of these molecules to antioxidants involved in protection of the cells (Toivonen, 1997, Keles and Oncel, 2002 and Vallejo et al., 2003).

Losses of L-ascorbic acid were lower in LCS broccoli (Fig. 1c) indicating that low SWC and CS synergistically contributed to its preservation. L-ascorbic acid reduction involves participation in the protection of the oxidative and respiratory systems, as well as in the recycling of lipid-soluble α-tocopherol synthesized in the mitochondria and transported to the chloroplasts, important organelles in oxidative stress protection (Smirnoff, 2000 and Serrano et al., 2006).

Preservation of 5-mthf in LCS broccoli could be associated with the preservation of L-ascorbic acid. Although L-ascorbic acid is mainly synthesized in the mitochondria, higher concentrations are found in the plastids where folates are synthesized. Alhagdow et al. (2007) also mentioned that L-ascorbic acid acting as an electron receptor in enzymatic and chemical oxidation–reduction reactions contributed to the preservation of folates, carotenoids, α-tocopherol and chlorophyll.

Antioxidant activity was highly correlated with L-ascorbic acid (0.84), 5-mthf (0.79) and carotenoids (0.63), and consequently, LCS broccoli with higher contents of these compounds had high antioxidant activity (Table 2). Often antioxidant activity in vegetables is strongly associated with phenolic compounds (Starzynska et al., 2003, Serrano et al., 2006 and Koca and Karadeniz, 2008). In this study, this was not the case, possibly due to the fact that contents of phenolic compounds were relatively low when compared to other genotypes (Zhang and Hamauzu, 2004).

In summary, the preservation of colour, L-ascorbic acid, 5-mthf and antioxidant activity was achieved as a consequence of the interaction between low SWC during cultivation and CS after harvest. On the other hand, losses of phenolic compounds were not dependent on SWC or PSC and carotenoid stability was only influenced by PSC.

ACKNOWLEDGEMENTS

To CNPq for financial support (project number 470480/2007-6) and a scholarship given to Cogo S.L.P.

REFERENCES

1. Able, A.J., Wong, L.S., Prasad, A., O'Hare, T.J., 2002. 1-MCP is more effective on a floral brassica (Brassica oleracea var. italica L.) than a leafy brassica (Brassica rapa var. chinensis). Postharvest Biol. Technol. 26, 147–155.

2. Able, A.J., Wong, L.S., Prasad, A., O'Hare, T.J., 2005. The physiology of senescence in detached pak choy leaves (Brassica rapa var. chinensis) during storage at different temperatures. Postharvest Biol. Technol. 35, 271–278.

3. Alhagdow, M., Mounet, F., Gilbert, L., Nunes-Nesi, A., Garcia, V., Just, D., Petit, J., Beauvoit, B., Fernie, A.R., Rothan, C., Baldet, P., 2007. Silencing of the mitochondrial ascorbate synthesizing enzyme l-

galactono-1,4-lactone dehydrogenase affects plant and fruit development in tomato. Plant Physiol. 145, 1408–1422.

4. Aluru, M.R., Zola, J., Foudree, A., Rodermel, S.R., 2009. Chloroplast photooxidationinduced transcriptome reprogramming in arabidopsis immutans white leaf sectors. Plant Physiol. 150, 904–923.

5. Asoda, T., Terai, H., Kato, M., Suzuki, Y., 2009. Effects of postharvest ethanol vapor treatment on ethylene responsiveness in broccoli. Postharvest Biol. Technol. 52, 216–220.

6. Brand-Williams, W., Cuvelier, M.E., Berset, C., 1995. Use of a free-radical method to evaluate antioxidant activity. LWT – Food Sci. Technol. 28, 25–30.

7. Chen, L.F.O., Hwang, J.Y., Charng, Y.Y., Sun, C.W., Yang, S.F., 2001. Transformation of broccoli (Brassica oleracea var. italica) with isopentenyltransferase gene via Agrobacterium tumefaciens for post-harvest yellowing retardation. Mol. Breed. 7, 243–257.

8. Cisneros-Zevallos, L., 2003. The use of controlled postharvest abiotic stresses as a tool for enhancing the nutraceutical content and adding-value of fresh fruits and vegetables. J. Food Sci. 68, 1560–1565.

9. Costa, L., Vicente, A.R., Civello, P.M., Chaves, A.R., Martínez, G.A., 2006. UV-C treatment delays postharvest senescence in broccoli florets. Postharvest Biol. Technol. 39, 204–210.

10. Costa, M.L., Civello, P.M., Chaves, A.R., Martinez, G.A., 2005. Effect of ethephon and 6- benzylaminopurine on chlorophyll degrading enzymes and a peroxidase-linked chlorophyll bleaching during post-harvest senescence of broccoli (Brassica oleracea L.) at 20 ◦C. Postharvest Biol. Technol 35, 191–199.

11. Eason, J.R., Patel, D., Ryan, D., Page, B., Hedderley, D., Watson, L., West, P., 2007. Controlled atmosphere treatment of broccoli after harvest delays senescence and induces the expression of novel BoCAR genes. Plant Physiol. Biochem. 45, 445–456.

12. Fortier, E., Desjardins, Y., Tremblay, N., Bélec, C., Côté, M., 2010. Influence of irrigation and nitrogen fertilization on broccoli polyphenolics concentrations. Acta Hort. 856, 55–62.

13. Funamoto, Y., Yamauchi, N., Shigyo, M., 2003. Involvement of peroxidase in chlorophyll degradation in stored broccoli (Brassica oleracea L.) and inhibition of the activity by heat treatment. Postharvest Biol. Technol 28, 39–46.

14. Hortensteiner, S., 2006. Chlorophyll degradation during senescence. Annu. Rev. Plant Biol. 57, 55–77.

15. Jones, R.B., Faragher, J.D., Winkler, S., 2006. A review of the influence of postharvest treatments on quality and glucosinolate content in broccoli (Brassica oleracea var. italica) heads. Postharvest Biol. Technol. 41, 1–8.

16. Keles, Y., Oncel, I., 2002. Response of antioxidative defence system to temperature and water stress combinations in wheat seedlings. Plant Sci. 163, 783–790.

17. Koca, N., Karadeniz, F., 2008. Changes of bioactive compounds and anti-oxidant activity during cold storage of carrots. Int. J. Food Sci. Technol. 43, 2019–2025.

18. Leja, M., Mareczek, A., Starzynska, A., Rozek, S., 2001. Antioxidant ability of broccoli flower buds during short-term storage. Food Chem. 72, 219–222.

19. Lichtenthaler, H.K., 1987. Chlorophylls and carotenoids – pigments of photosynthetic biomembranes. Methods Enzymol. 148, 350–382.

20. Mittler, R., 2002. Oxidative stress, antioxidants and stress tolerance. Trends Plant Sci. 7, 405–410.

21. Moreno, D.A., Carvajal, M., López-Berenguer, C., García-Viguera, C., 2006. Chemical and biological characterisation of nutraceutical compounds of broccoli. J. Pharm. Biomed. Anal. 41, 1508–1522.

22. Oh, M.-M., Trick, H.N., Rajashekar, C.B., 2009. Secondary metabolism and antioxidants are involved in environmental adaptation and stress tolerance in lettuce. J. Plant Physiol. 166, 180–191.

23. Podsedek, A., 2007. Natural antioxidants and antioxidant capacity of Brassica vegetables: a review. LWT – Food Sci. Technol. 40, 1–11.

24. Pruzinska, A., Tanner, G., Aubry, S., Anders, I., Moser, S., Muller, T., Ongania, K.H., Krautler, B., Youn, J.Y., Liljegren, S.J., Hortensteiner, S., 2005. Chlorophyll breakdown in senescent Arabidopsis leaves. Characterization of chlorophyll catabolites and of chlorophyll catabolic enzymes involved in the degreening reaction. Plant Physiol. 139, 52–63.

25. Rai, D.R., Tyagi, S.K., Jha, S.N., Mohan, S., 2008. Qualitative changes in the broccoli (Brassica oleracea Italica) under modified atmosphere packaging in perforated polymeric films. J. Food Sci. Technol. 45, 247–250.

26. Rakotonirainy, A.M., Wang, Q., Padua, G.W., 2001. Evaluation of zein films as modi- fied atmosphere packaging for fresh broccoli. J. Food Sci. 66, 1108–1111.

27. Schelbert, S., Aubry, S., Burla, B., Agne, B., Kessler, F., Krupinska, K., Hortensteiner, S., 2009. Pheophytin pheophorbide hydrolase (pheophytinase) is involved in chlorophyll breakdown during leaf senescence in arabidopsis. Plant Cell 21, 767–785.

28. Serrano, M., Martinez-Romero, D., Guillen, F., Castillo, S., Valero, D., 2006. Maintenance of broccoli quality and functional properties during cold storage as affected by modified atmosphere packaging. Postharvest Biol. Technol. 39, 61–68.

29. Smirnoff, N., 2000. Ascorbate biosynthesis and function in photoprotection. Philos. Trans. R. Soc. Lond. Ser. B: Biol. Sci. 355, 1455–1464.

30. Starzynska, A., Leja, M., Mareczek, A., 2003. Physiological changes in the antioxidant system of broccoli flower buds senescing during short-term storage, related to temperature and packaging. Plant Sci. 165, 1387–1395.

31. Synkova, H., Semoradova, S., Schnablova, R., Witters, E., Husak, M., Valcke, R., 2006. Cytokinin-induced activity of antioxidant enzymes in transgenic Pssu-ipt tobacco during plant ontogeny. Biol. Plant. 50, 31–41.

32. Toivonen, P.M.A., 1997. The effects of storage temperature, storage duration, hydrocooling, and micro-perforated wrap on shelf life of broccoli (Brassica oleracea L., Italica Group). Postharvest Biol. Technol. 10, 59–65.

33. Vallejo, F., Tomas-Barberan, F., Garcia-Viguera, C., 2003. Health-promoting compounds in broccoli as influenced by refrigerated transport and retail sale period. J. Agric. Food Chem. 51, 3029–3034.

34. Watada, A.E., Ko, N.P., Minott, D.A., 1996. Factors affecting quality of fresh-cut horticultural products. Postharvest Biol. Technol. 9, 115–125.

35. Wurr, D.C.E., Hambidge, A.J., Fellows, J.R., Lynn, J.R., Pink, D.A.C., 2002. The influence of water stress during crop growth on the postharvest quality of broccoli. Postharvest Biol. Technol. 25, 193–198.

36. Xu, Y., Huang, B.R., 2009. Effects of foliar-applied ethylene inhibitor and synthetic cytokinin on creeping bentgrass to enhance heat tolerance. Crop Sci. 49, 1876–1884.

37. Zaicovski, C.B., Zimmerman, T., Nora, L., Nora, F.R., Silva, J.A., Rombaldi, C.V., 2008. Water stress increases cytokinin biosynthesis and delays postharvest yellowing of broccoli florets. Postharvest Biol. Technol. 49, 436–439.

38. Zhang, D.L., Hamauzu, Y., 2004. Phenolics, ascorbic acid, carotenoids and antioxidant activity of broccoli and their changes during conventional and microwave cooking. Food Chem. 88, 503–509.

Chapter 6

Morphological and molecular characterization of ethylene binding inhibition in carnations

B.C. In[a], J. Strable[a, b], B.M. Binder[c], T.G. Falbel[a], S.E. Patterson[a]

a Department of Horticulture, University of Wisconsin-Madison, Madison, WI 53706, USA

b Department of Genetics, Development and Cell Biology, Iowa State University, Ames, IA 50011, USA

c Department of Biochemistry, Cellular, and Molecular Biology, University of Tennessee, Knoxville, TN 37996, USAa

HIGHLIGHTS

- Ethylene responses in carnation.
- Ethylene biosynthesis in carnation.
- Molecular responses to ethylene inhibition with 1-MCP.
- Molecular responses to loss of ethylene inhibition with 1-MCP.
- Increased understanding of ethylene and 1-MCP response in cut flowers.

ABSTRACT

Ethylene perception by ethylene receptors can be suppressed by ethylene antagonists, such as 1-methylcyclopropene (1-MCP). 1-MCP binds to ethylene receptors blocking ethylene binding, and thereby represses ethylene responses, including flower senescence and petal abscission. Despite its antagonistic propensity, plants treated with 1-MCP often regain sensitivity to ethylene, suggesting that ethylene receptors are synthesized de novo post-treatment. To investigate this observation, we determined the relationship between the mRNA levels of ethylene biosynthesis and ethylene receptor genes and the degree of ethylene sensitivity of carnation flowers after ethylene and 1-MCP treatments. Flowers treated with a single application of 1-MCP lost the inhibitory effect in ethylene production after 7 d of the treatment. In contrast, multiple treatments with 1-MCP completely suppressed several ethylene biosynthesis genes and ethylene production throughout the experiment. Multiple 1-MCP treatments increased vase life by almost three-fold relative to control flowers. Eventually ethylene-independent browning and desiccation on the edges of petals contributed to loss of vase life. We observed that the transcript levels of the ethylene receptor genes, *DcETR1* and *DcERS1*, decreased with flower development, but increased at the onset of floral senescence. Our results suggest that ethylene receptors are continually synthesized during later stages of floral development. While ethylene responsiveness is temporarily blocked by 1-MCP treatment, we predict that new receptors synthesized days after 1-MCP treatment can bind ethylene. This leads to the observed recovery of ethylene sensitivity in the flowers. In addition, we predict that the degradation of ethylene receptors that is stimulated by ethylene binding is prevented by successive treatments with 1-MCP prior to recovery of ethylene sensitivity. This study improves our understanding about the uses and effectiveness of 1-MCP and will facilitate the development and improvement of postharvest treatments for multiple crops.

KEYWORDS

Carnation; Ethylene biosynthesis; Ethylene receptor; Flower senescence; 1-MCP; Cut flowers

1. INTRODUCTION

Cut flower longevity has been a concern of growers for decades. Although there are numerous practices and treatments to prolong the vase life of cut flowers, the process of flower senescence is still poorly understood. The plant hormone ethylene regulates multiple aspects of plant growth and development, including seed germination, abscission, flower senescence, and fruit ripening (Abeles et al., 1992 and Reid and Wu, 1992). Ethylene is synthesized throughout the plant via a well-characterized biosynthetic pathway (Yang and Hoffman, 1984 and Kende, 1993).

In many flowers, the onset of flower senescence is associated with a climacteric increase in ethylene biosynthesis. Once the autocatalytic synthesis of ethylene is initiated, the senescence process is considered irreversible in ethylene-sensitive flowers such as carnations (Halevy and Mayak, 1981 and Yang and Hoffman, 1984). Ethylene perception by ethylene receptors is an indispensable requirement to commence and sustain the ethylene-mediated senescence program (Borochov and Woodson, 1989). When ethylene is perceived by the receptors, the ethylene signal is sent through a sequence of biochemical events that regulate the expression of ethylene responsive genes leading to ethylene production and ultimately flower senescence. The ability to perceive or respond to ethylene is most likely mediated by changes in ethylene signaling during flower development (Halevy and Mayak, 1981, Woodson and Lawton, 1988 and Verlinden et al., 2002).

From multiple genetic studies in the model plant *Arabidopsis thaliana*, a family of ethylene receptors (ETR1, ETR2, ERS1, ERS2 and EIN4), as well as several downstream genes (CTR1, EIN2, EIN3, EIL1, and ERF1) have been discovered that act in a signal transduction cascade (Schaller, 2012). Ethylene perception by the family of receptors is thought to inactivate the antagonist CTR1, a Raf-like MAP kinase kinase kinase (MAPKKK), thereby repressing the negative regulation of these receptors at the cell membrane (Chang et al., 1993, Kieber et al., 1993, Hua and Meyerowitz, 1998, Hua et al., 1998 and Sakai et al., 1998).

Historically, it has been shown that the ethylene receptor isoforms have ethylene binding sites with similar binding affinities for ethylene (Schaller and Bleecker, 1995, Hall et al., 2000 and O'Malley et al., 2005). Ethylene binding and action can be suppressed by the potent inhibitor 1-methylcyclopropene (1-MCP). Interestingly, 1-MCP has been shown to bind competitively to the ethylene receptors with a greater affinity for the receptors than that of ethylene (Sisler et al., 1986, Sisler et al., 1990 and Hall et al., 2000). Therefore, it has been proposed that 1-MCP can block the ethylene binding to the receptor and inhibit down-stream components in ethylene signal transduction. Consequently, it was found that 1-MCP works effectively in many horticultural plants to prolong their postharvest longevity such as carnation (Sisler et al., 1996), *Alstroemeria*, *Kalanchoe* (Serek et al., 1994 and Serek et al., 1995), *Pelargonium* (Cameron and Reid, 2001), banana, avocado and apple (Jiang et al., 1999, Watkins et al., 2000, Blankenship and Dole, 2003 and Schotsmans et al., 2009). In addition, it has been shown that 1-MCP is effective at very low concentration (10–50 nL L^{-1} for 6 h) in carnation flowers (Serek et al., 1995,Sisler et al., 1996 and Reid and Çelikel, 2008).

1-MCP is hypothesized to be bound to ethylene receptors irreversibly *in planta*; and thus, plants treated with 1-MCP should not respond to ethylene (Serek et al., 1994, Sisler et al., 1996, Sisler and Serek, 2003 and Binder et al., 2004a). Despite this irreversible binding to the receptors, plants treated with 1-MCP often regain sensitivity to ethylene (i.e., recover) post-treatment. However, plants that undergo successive 1-MCP treatments prior to their recovery of ethylene sensitivity have prolonged repression of ethylene responses (Cameron and Reid, 2001 and Ekman et al., 2004).

Based on an irreversible antagonist model for 1-MCP action, it has been proposed that the recovery of ethylene sensitivity in plants after 1-MCP treatment is the consequence of synthesis of new ethylene receptors during development and senescence of the flowers (Binder and Bleecker, 2003 and Schotsmans et al., 2009). However, to our knowledge, no data directly supports this hypothesis and it is not understood why some plants respond differently than others. In this study, we set out to test the hypothesis that ethylene receptors are synthesized de novo, and that this accounts for the ability of the carnation plants to regain ethylene sensitivity post-treatment with 1-MCP.

Carnation (*Dianthus caryophyllus* L.) flowers were selected for this study due to their high sensitivity and rapid response to gaseous regulators such as ethylene and 1-MCP. Three ethylene receptor genes have been identified in carnation plants, *DcETR1*, *DcERS1* and*DcERS2* that are orthologous to the ethylene receptors of *Arabidopsis* (Charng et al., 1997, Shibuya et al., 1998 and Nagata et al., 2000). We took a molecular approach to explore how plants regain the sensitivity to ethylene at the transcript level by monitoring the relationship between mRNA levels of ethylene biosynthesis genes and receptor genes, and the sensitivity of flowers to ethylene at various stages of development after single or multiple 1-MCP treatments. Floral morphology was also observed before and after treatments to determine the effects of 1-MCP and ethylene on carnation flowers.

2. MATERIALS AND METHODS

2.1. Plant materials

Rooted cuttings of carnation plants (*Dianthus caryophyllus* L. cv. Moonstone) were planted in 15 cm plastic pots with growing medium (Metro-Mix Special Blend, SUNGRO Horticulture Distribution Inc., Bellevue, WA). The plants were grown on benches in a greenhouse at 22/16 °C day/night temperature and flood-irrigated with half strength Hoagland nutrient solution. Supplementary lighting (220 µmol m^{-2} s^{-l} PPFD at plant level) was provided by high-pressure sodium lamps (Philips 600 W Master GreenPower) from 05:00 until 21:00 to assure a photoperiod of 16 h. These lamps were turned on and off automatically by a light sensor located inside the ceiling of the greenhouse that operated when the intensity of natural lighting was less than or exceeded 350 µmol m^{-2} s^{-l}. Plants were pinched 4 weeks after planting. Carnation flowers were harvested when the outer petals were just horizontal and immediately placed in tap water for all experiments.

2.2. Treatment with ethylene and 1-MCP

Flowers were trimmed to 5 cm stem length, and placed in 20 mL glass vials containing distilled water. The flowers were subsequently treated singly with air, 5 µL L^{-1} ethylene, 100 nL L^{-1} 1-MCP for 6 h; or repeated 1-MCP treatments every other day. Air treated flowers were used as controls. Control and treated flowers were enclosed in dark treatment chambers (117 L) with air circulation by a small fan. Control flowers were maintained in the chamber with normal atmospheric air. For ethylene treatment, 5.85 mL of 10% ethylene was injected into the chamber to give a final concentration of 5 µL L^{-1}. The ethylene action inhibitor, 1-MCP was applied via SmartFresh (AgroFresh Inc., Philadelphia, PA) tablets placed in the treatment chambers at room temperature. To generate 100 nL L^{-1} 1-MCP gas, a gray research tablet (0.026 mg 1-MCP) and an activator tablet was added to a beaker containing 18 mL of activator solution and was placed in the chamber. The cut flowers were transferred to the chamber along with small fan to provide air circulation during the treatments (ethylene or air) and then the chamber was tightly sealed for 6 h. After the treatments, the flowers were held in environmental controlled room at 20 ± 2 °C, 40 ± 5% relative humidity and about 50 μmol m^{-2} s^{-1} of fluorescent lighting for 12 h.

2.3. Ethylene measurements

For ethylene measurements, three carnation flowers in each treatment were selected at four distinct stages of flower development: day 1, flowers beginning to open; day 3, flowers completely opened; day 6, flowers showing onset of petal in-rolling; and day 7, flowers with the petal in-rolled. Individual flowers were placed in a 500 mL glass jar with a rubber septum and were enclosed for 1 h at room temperature. A 1 mL gas sample was collected with a gas-tight hypodermic syringe through the septum and analyzed for ethylene by a gas chromatograph (Model 8500, PerkinElmer Corp., Norwalk, CT) equipped with an aluminum column and flame ionization detector. Each experiment was repeated at least three times.

2.4. cDNA synthesis and semi-quantitative RT-PCR

Receptacles, petal claw and petal blade were used for the gene expression analysis since the floral organs show a distinct differences in the responsiveness to ethylene (Overbeek and Woltering, 1990 and Drory et al., 1993). After the assay for ethylene production, petals and the receptacle were detached from the flowers and the petals were divided into upper (petal blade) and basal (petal claw) portions (Mor et al., 1985). For tissue-specific expression profiles, petals, ovaries, receptacles, styles, stamens, and sepals were also detached from fully opened flowers. Immediately after detachment, they were frozen in liquid nitrogen and stored at −80 °C for RNA isolation. Total RNA was isolated from the various floral organs using Trizol method according to the manufacturer's procedure with slight modifications. Approximately 300 mg of tissues were ground in the presence of liquid nitrogen to a fine powder using a pre-chilled mortar and pestle and were homogenized in 1 mL Trizol Reagent (Invitrogen, Carlsbad, CA). The sample in Trizol was separated into aqueous and hydrophobic phases with 200 μL chloroform by centrifugation. The RNA was precipitated from the aqueous phase with 50 µL of 5 M NaCl and 0.5 mL isopropanol by centrifugation, washed with 80% ethanol, and resuspended in RNase-free H_2O. Total RNA was quantified with NanoDrop DN-1000 spectrophotometer (NanoDrop Technologies Inc., Wilmington, DE). RNA samples were treated with RNase-free DNase prior to RT-PCR and first-strand cDNA was synthesized from 2 µg of total RNA with 1 µg of oligo $(dT)_{18}$ primer, dNTPs, RNase inhibitor, buffer, and M-MLV reverse transcriptase in a final volume of 25 µL according to the manufacturer's instructions (Promega, WI). Reverse transcription was performed in a PTC-200 PCR machine (MJ Research Inc., MA) with the following temperature parameters: 15 min at 70 °C followed by 1 h at 42 °C.

To determine the expression patterns of ethylene biosynthesis and receptor genes during flower senescence, a semi-quantitative RT-PCR analysis of the transcript levels was performed using gene-specific primers. These experiments were performed with three independent biological replicates. Based on database published sequences, a pair of specific primers was designed for *DcACS1*, *DcACO1*, *DcETR1*, *DcERS1* and *DcERS2*genes and synthesized by Integrated DNA Technologies (Coralville, IA). Carnation actin (*DcACT1*) was used as an internal control. Primer pairs that were used for the semi-quantitative PCR analyses are listed in Table 1. For PCR reactions, 1 µL of cDNA was used as a template with 2 µL of PCR buffer, 0.2 µL of Taq polymerase, dNTPs, and forward and reverse primers in a final volume of 20 µL. PCR amplification was performed with the following temperature parameters: 95 °C for 3 min and then cycled 35 (*DcERS2)*or 40 times (all the other genes) at 95 °C for 20 s, 60 °C for 30 s, and 72 °C for 40 s, with a final extension of 7 min at 72 °C. The PCR products were then analyzed by 1% (w/v) agarose gel and stained with ethidium bromide. The gels were visualized under UV light and the images were taken using a gel documentation system (Gel Doc XR+, Bio-Rad, Hercules, CA) and quantified with ImageJ 1.43u (National Institutes of Health, USA). The relative level of the band was showed as the absolute integrated absorbency normalized to the relative actin band.

Table 1. Gene-specific primers used for RT-PCR amplification of cDNA fragments.

Gene (accession number)	Forward primer	Reverse primer	Cycles
DcACS1 (M66619)	5′-AAAATCGCTACGAACGATGG-3′	5′-TGCGGGATAATAAGGAGACG-3′	40
DcACO1 (AB042320)	5′-TCACAACTGGGGATTCTTCC-3′	5′-CATCTGTCTGCGCTATCACG-3′	40
DcETR1 (AB035806.1)	5′-CACACTCCGTCAGCAATATCC-3′	5′-TCCATAAGCTGTTGCTGTGC-3′	40
DcERS1 (AF016250.1)	5′-CGTCTGTGCTGTAGGTGACG-3′	5′-CAAATTGCGAGTCCAAGTCC-3′	40
DcERS2 (AF034770.1)	5′-AGAGTGACGGTCATGACAAGG-3′	5′-CCTAACCGCGCTTAACTCC-3′	35
DcACT1 (AY007315)	5′-GGCCGTTCTCTCTCTGTACG-3′	5′-GTCCATCAGGCATTTCATAGC-3′	35

3. RESULTS

3.1. Suppression of ethylene biosynthesis in floral organs by 1-MCP treatment

To understand the relationship between ethylene biosynthesis and flower senescence, the expression of ethylene biosynthesis genes, ethylene production, and flower longevity were determined at four stages of flower development after treatment with ethylene, a single application of 1-MCP, or multiple applications of 1-MCP. The expression levels of ethylene biosynthesis genes *DcACS1* and *DcACO1* were monitored during flower development and senescence (Fig. 1). The transcript levels of *DcACS1* and *DcACO1* in control flowers were very low or undetectable until day 3 and increased significantly in all floral organs during late senescent stages (days 6 and 7). In response to ethylene treatment, initial transcript levels of *DcACO1* increased significantly in all tissues (Fig. 1B) while changes in levels of *DcACS1* were not discernable (Fig. 1A). These findings enhance previous observations on spatial and temporal specificity of ethylene biosyntheis genes in carnation petals. Previous studies reported differences in responsiveness of these ethylene biosynthesis genes indicating that the expression levels of *DcACS1* are similar or more abundant than *DcACO1* during natural senescence and that *DcACO1* is rapidly induced by ethylene treatment (Woodson et al., 1992 and ten Have and Woltering, 1997). We also observed that a single treatment with 1-MCP completely inhibited the changes in transcript levels of *DcACS1* in specific tissues including petal claw and petal blade. However, there was no measurable effect on *DcACS1* expression in the receptacle.

We determined that ethylene production in whole flowers primarily mirrored the expression patterns of the ethylene biosynthesis genes *DcACS1* and *DcACO1* (Fig. 1 and Fig. 2). Production in untreated control flowers showed a dramatic increase at day 6, followed by a decrease at day 7. Flowers treated with ethylene produced 30% higher levels of ethylene. In contrast, flowers treated with a single application of 1-MCP displayed a delayed increase in ethylene production that was not measurable until day 7. These plants produced approximately 50% lower levels of ethylene than control plants. Flowers treated with multiple applications of 1-MCP had no detectable ethylene production.

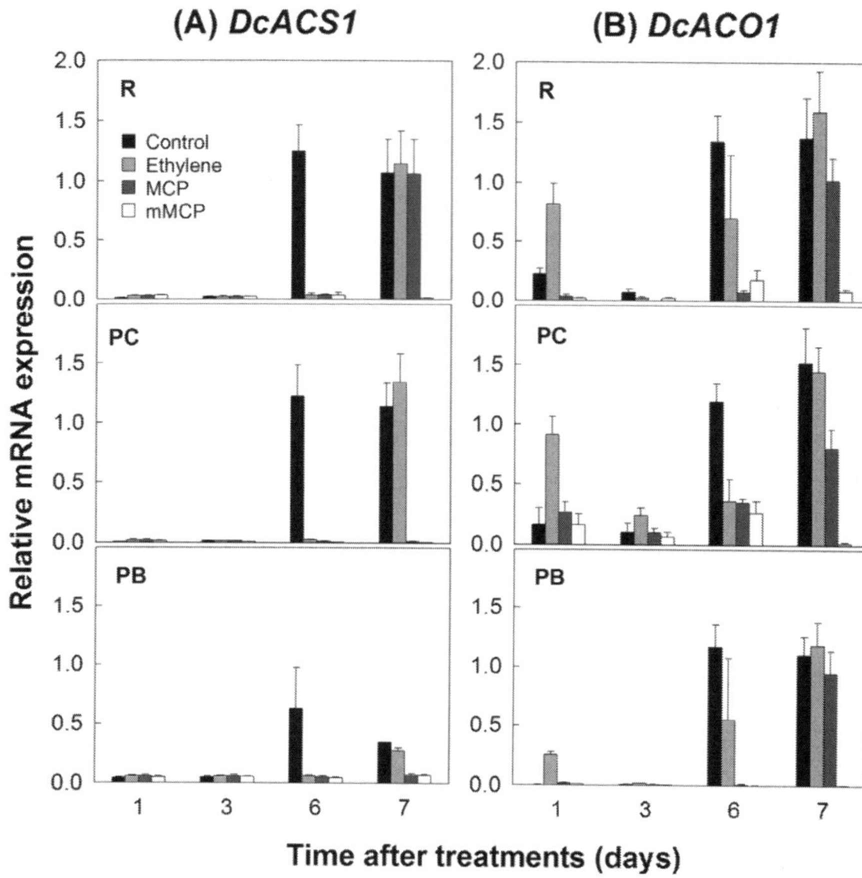

Figure 1. Expression patterns of *DcACS1 and DcACO1* genes in floral organs of carnations. Flowers were harvested at the fully opened stage and immediately treated singly with ethylene (5 μL L^{-1}); 1-MCP (100 nL L^{-1}) for 6 h; or multiple 1-MCP treatments every other day. Total RNA was isolated from the receptacle (R), petal claw (PC), and petal blade (PB) immediately after the determination of ethylene production in whole flowers. Semi-quantitative RT-PCR analysis was performed to determine the mRNA level of *DcACS1* and *DcACO1*. The relative expression level was normalized to actin (*DcACT1*). Control, Air only; MCP, single 1-MCP treatment; mMCP, multiple 1-MCP treatments at day 1, day 3, day 5 and day 7. Data represents the mean ± SE of three replicates.

Flower longevity was strongly correlated with ethylene production patterns. Both control and ethylene-treated flowers demonstrated obvious signs of senescence by day 7 with more pronounced symptoms in the ethylene treated flowers. Specific morphological changes included petal inrolling and wilting, and these flowers completely ended their vase life on day 8 due to these changes (Fig. 2B and C). Vase life of the flowers after a single 1-MCP treatment was prolonged by about 2 d relative to control flowers. When the flowers were treated with multiple applications of 1-MCP, vase life was extended by 14 d compared to control flowers (Fig. 2C). The flowers treated with multiple applications of 1-MCP did not show petal inrolling during the vase period. However, they eventually lost their vase life after 20 d due to browning or desiccation around the petal margins (Fig. 2D); a senescence symptom that occurs in many ethylene-insensitive floral organs (Tanase et al., 2008).

3.2. Tissue-specific expression profiles of ethylene biosynthesis and receptor genes

In order to understand ethylene responses in carnation flowers more fully, we also looked at the relationship of tissue specific expression of ethylene biosynthesis and receptor genes in the different floral organ tissues using RT-PCR. The expression profiles of ethylene biosynthesis and receptor genes revealed tissue-specific signatures in floral organs of carnations at the fully opened stage (Fig. 3). The ethylene biosynthesis gene*DcACS1* was not detected in most of the tissues with the exception of weak expression in the ovary. Alternately, *DcACO1* transcript was abundant in the receptacle, ovary, stigma and stamen. However, *DcACO1* expressed very weakly in the petal claw, and was not detected in the petal blade of flowers at the fully opened stage. Transcripts for all three ethylene receptor genes were found in all tissues studied; although, the patterns of expression varied for each receptor isoform. *DcETR1* transcript levels were generally lower than the other isoforms. Alternatively, *DcERS1* transcripts were more abundant in the receptacle, ovary and style, and less abundant in the petal and stamen. However,*DcERS2* transcripts generally showed equal abundance in all floral organs except the petal claw where there was less transcript (Fig. 3).

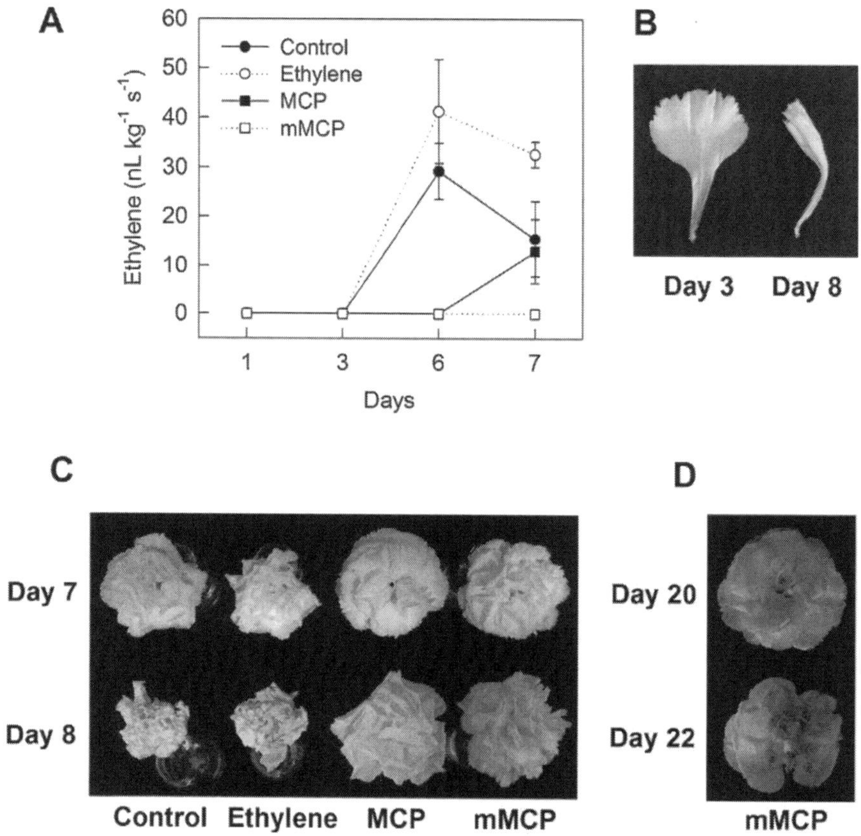

Figure 2. Effects of ethylene and 1-MCP treatment on ethylene production and flower senescence during vase life in carnation cv. Moonstone. (A) Changes in ethylene production of whole flowers; (B) carnation petal from a fully open flower (Day 3) and senescent flower showing inrolling (Day 8); (C) flower senescence on day 7 and day 8 after treatments with ethylene or 1-MCP; (D) flower senescence on day 20 and day 22 in the flowers treated multiple times with 1-MCP. Flowers were harvested at the fully opened stage and immediately treated singly with ethylene (5 μL L^{-1}); 1-MCP (100 nL L^{-1}) for 6 h; or multiple 1-MCP treatments every other day. Control, Air only; MCP, 1-MCP single; mMCP, multiple 1-MCP treatments at day 1, day 3, day 5 and day 7. Data represents the mean ± SE of three replicates.

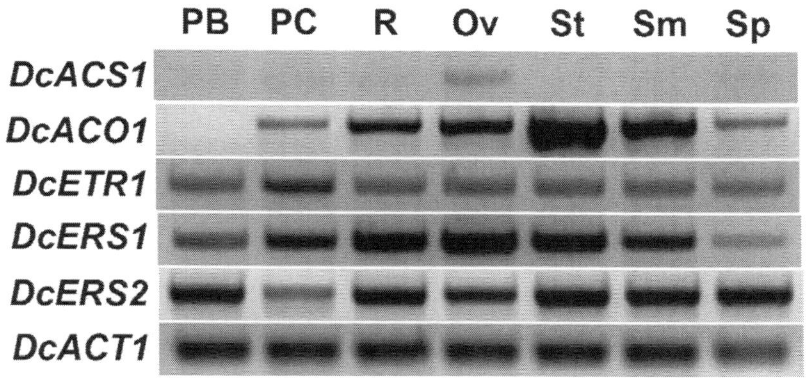

Figure 3. Tissue-specific expression profiles of *DcACS1, DcACO1, DcETR1, DcERS1* and *DcERS2* genes in carnation flowers. Expression analysis was performed on various floral organs of carnations at the fully opened stage using RT-PCR. PB, petal blade; PC, petal claw; R, receptacle; Ov, ovary; St, style; Sm, stamen; Sp, sepal.

3.3. Expression pattern of ethylene receptor genes during development and senescence of flowers

To further assess the role of ethylene receptor genes in ethylene sensitivity, the expression patterns of the three ethylene receptor genes, *DcETR1*, *DcERS1* and*DcERS2* were determined in the receptacle, petal claw and petal blade at various stages of flower development after treatments with ethylene or 1-MCP (Fig. 4). The expression of *DcETR1* in air control treatments was fairly constitutive in all tissues with a slight initial decrease in expression that might have been in response to recovery from harvest. We observed that *DcETR1* expression in the petal claw and petal blade was only slightly affected by either ethylene or 1-MCP treatments when compared with the controls. However, in the receptacle, *DcETR1* transcript accumulation was affected by both ethylene and 1-MCP treatments throughout floral development. Subsequent to ethylene treatment, *DcETR1* transcript levels in the receptacle increased more than 2-fold relative to control at the later stages of flower development. While *DcETR1* expression was markedly increased by day 7 after ethylene treatment, its expression was suppressed by multiple applications of 1-MCP. We also observed slight increases in *DcETR1* transcript at days 6 or 7 in the receptacle for all treatments except the multiple 1-MCP treatments (Fig. 4A).

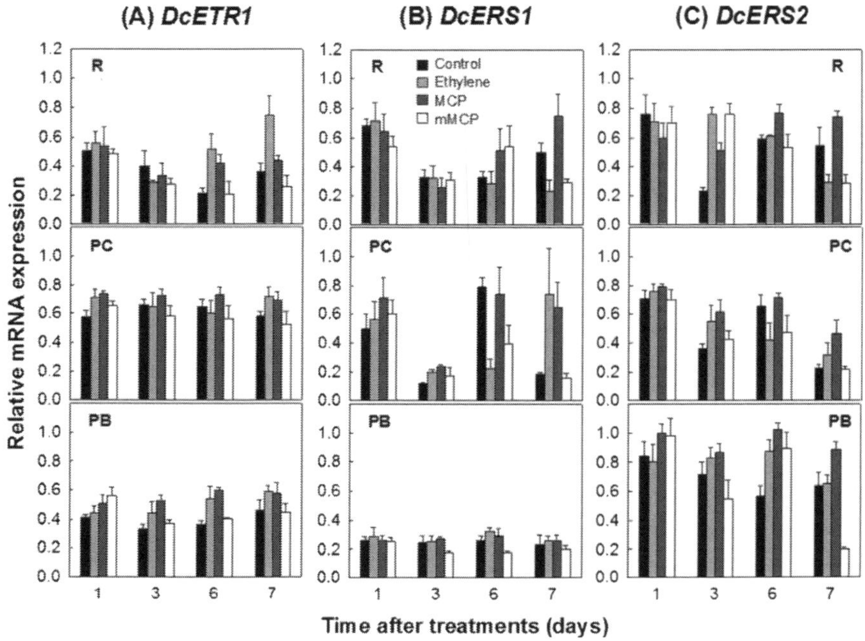

Figure 4. Expression patterns of *DcETR1, DcERS1* and *DCERS2* genes in floral organs of carnations. Flowers were harvested at the fully opened stage and immediately treated singly with ethylene (5 µL L^{-1}); 1-MCP (100 nL L^{-1}) for 6 h; or multiple 1-MCP treatments every other day. Total RNA was isolated from receptacle (R), petal claw (PC), and petal blade (PB) immediately after the determination of ethylene production in whole flowers as shown in Fig. 1. Semi-quantitative RT-PCR analysis was performed to determine the mRNA level of *DcETR1, DcERS1* and *DCERS2*. The relative expression level was normalized to actin (*DcACT1*). Control, Air only; MCP, single 1-MCP treatment; mMCP, multiple 1-MCP treatments at day 1, day 3, day 5 and day 7. Data represents the mean ± SE of three replicates.

In general, the expression level of *DcERS1* mRNA dropped rapidly at the fully open stage of flower development on day 3 and then showed increasing levels toward the late senescent stages in both receptacle and petal claw. Control tissues showed a consistently high level of *DcERS1* expression in petal blade throughout flower development and senescence. Following ethylene treatment, *DcERS1* transcript level increased considerably by day 7 in petal claw, but remained lower than untreated control in receptacle throughout flower development. In petal blades, *DcERS1 expression* was not clearly affected by ethylene treatment compared to control. Curiously, single applications of 1-MCP had no measurable effect on *DcERS1* expression in all floral organs. Overall, multiple treatments with 1-MCP suppressed changes in *DcERS1* transcript levels at the later stages of floral development. The mRNA levels were lowered by multiple applications of 1-MCP compared to control and ethylene-treated tissues. This was especially apparent in the petal tissues (Fig. 4B).

The mRNA levels of *DcERS2* in control plants initially showed higher accumulation and subsequently a rapid decrease in fully opened flowers. The *DcERS2* transcript in the receptacle increased in association with flower senescence at the later stages. In response to ethylene, *DcERS2* expression was elevated in receptacle tissues at the fully opened stage of flower development and reduced at later stages. However, in petal blade tissues, the expression patterns of *DcERS2* appeared unaffected by ethylene when compared with the control. *DcERS2* expression in 1-MCP single-treated tissues showed similar patterns with that in control tissues in receptacle. *DcERS2* expression decreased on day 3 and then increased at the late senescent stages. However, multiple treatments with 1-MCP resulted in decreased levels of expression in petal blade and receptacle compared to control at the later stage on day 7 (Fig. 4C).

These transcript data indicate that both *DcETR1* and *DcERS1* expression is more significantly changed than *DcERS2* expression in response to ethylene treatment. Overall, the expression levels of receptor genes was enhanced by ethylene treatment, and inhibited by multiple treatments with 1-MCP, indicating that higher levels of expression correlate with floral senescence.

3.4. Changes in floral morphology and petal turgidity

To understand the relationship between the expression patterns of ethylene receptor genes, flower sensitivity to ethylene, and morphological changes, the changes in petal turgidity were investigated throughout flower development and senescence. Petal turgidity was observed in carnation flowers after all sepals were removed from the flowers to exclude the influence of calyx. Control flowers maintained turgid petals until the fully open flower stage of development (day 3) with obvious wilting at day 6, and complete loss of turgidity and inrolling on day 7. While limited morphological differences in petals were observed between control and ethylene-treated flowers, ethylene treatment caused the flowers to lose petal turgidity slightly earlier than the control flowers. Flowers that were treated with 1-MCP (both singly and multiple times) lacked inrolling throughout the experiment indicating reduced loss of turgidity (Fig. 5).

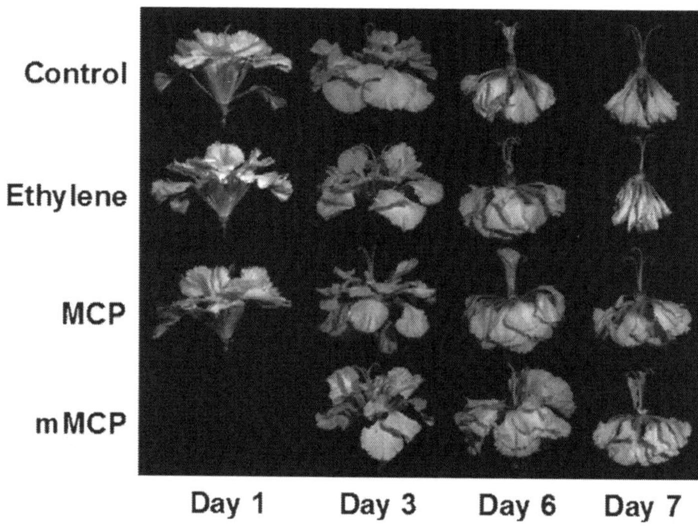

Figure 5. Changes in petal turgidity in carnation flowers during vase life. Flowers were harvested at the fully opened stage and immediately treated singly with ethylene (5 μL L^{-1}); 1-MCP (100 nL L^{-1}) for 6 h; or 1-MCP multiple treatments every the other day. Photographs were taken after removal of sepals from the flowers to exclude the influence of sepal. Control, Air only; MCP, single 1-MCP treatment; mMCP, multiple 1-MCP treatments at day 1, day 3, day 5 and day 7.

4. DISCUSSION

Ethylene perception by ethylene receptors can be suppressed by the competitive inhibitor 1-MCP. However, the inhibitory effects of 1-MCP on ethylene responses are often transient (Cameron and Reid, 2001 and Ekman et al., 2004). To better understand the antagonistic effects of 1-MCP, we compared the levels of specific genes, ethylene production and vase life in cut carnations after treatment with ethylene versus single and multiple treatments with 1-MCP. Single treatment of cut carnation flowers with ethylene resulted in measurable increases in autocatalytic ethylene production. This was paralleled by increases in the transcript levels of specific ethylene biosynthesis genes. In contrast, single treatments with 1-MCP resulted in delay in the onset of ethylene production and decreases in overall levels of ethylene production. Multiple 1-MCP treatments blocked increases in ethylene production. Overall, analysis of transcript levels of ethylene biosynthesis genes was consistent with these results.

We also examined the effects of 1-MCP on the vase life of cut carnations. Flowers treated with a single 1-MCP application had prolonged vase life characterized by a delay in petal inrolling. In contrast, flowers treated with multiple applications of 1-MCP, did not show petal inrolling during the vase period, but eventually lost their vase life after 20 d due to browning or desiccation around the petal margins. Interestingly, this is a common senescence symptom that occurs in ethylene-insensitive floral organs suggesting that the multiple 1-MCP treatments are completely blocking ethylene perception. These results are consistent with previous works showing that re-treatment with 1-MCP prior to regaining ethylene sensitivity effectively prevented the ethylene responses in*Pelargonium*s, waxflowers and pears (Cameron and Reid, 2001, Ekman et al., 2004 and Macnish et al., 2004). Together, these results indicate that cut carnations can regain sensitivity to ethylene after a single 1-MCP treatment, but that the effects of ethylene are completely blocked by multiple treatments with 1-MCP.

We also observed tissue-specific differences in gene expression and regulation. In higher plants, the ethylene biosynthetic pathway involves the conversion of S-adenosylmethionine (SAM) to 1-aminocyclopropane-1-carboxylic acid (ACC) by ACC synthase (ACS), and ACC is then converted to ethylene by ACC oxidase (ACO). ACS and ACO are generally regarded as rate-limiting enzymes in ethylene biosynthesis (Yang and Hoffman, 1984 and Kende, 1993). Presently, three ACS genes (*DcACS1, DcACS2* and*DcACS3*) and one ACO gene (*DcACO1*) have been identified in carnation (Park et al., 1992 and Henskens et al., 1994). *DcACS1* (*CARACC3*) has been shown to have a role in autocatalytic ethylene production in carnation flowers (ten Have and Woltering, 1997,Jones and Woodson, 1999, Jones, 2003 and Harada et al., 2011). In our examination of the transcript levels of ethylene biosynthesis genes in specific floral organ tissues we found that there were differences in the ethylene responsiveness of specific transcripts. Consistent with prior studies (Jones, 2003), we found that expression of the *DcACS1* and*DcACO1* transcripts were up-regulated slightly by ethylene and down-regulated by 1-MCP. A single treatment with 1-MCP prevented *DcACS1* transcript accumulation in petal claw and petal blade throughout the experimental period, but did not inhibit its expression levels in the receptacle at later stages pointing to key tissue-specific differences in the regulation of this gene. Interestingly, *DcACO1* transcript was transiently enhanced by exogenous ethylene on day 1 but did not result in increased ethylene production.

Transcripts for the receptors also showed interesting differences between the floral organs and the receptor isoforms. For instance, the transcript levels of *DcERS1* and*DcERS2* were much more abundant than that of *DcETR1* in the floral organ tissues of newly opened flowers suggesting these receptor isoforms have a larger role at this developmental stage. This is similar to the pattern observed in *Arabidopsis* where the*ETR1* and *ERS1* transcripts comprise more than 80% of total transcript levels of ethylene receptors (Binder et al., 2004b and O'Malley et al., 2005) and these receptor isoforms have a more pronounced role in ethylene signaling (Wang et al., 2003).

The accumulation of these transcripts was also influenced by ethylene and 1-MCP at a tissue-specific level. Especially, in receptacles, *DcETR1* transcript was significantly affected by both ethylene and multiple application of 1-MCP at the later stage, while in the petal tissues, it was only slightly influenced by either ethylene or 1-MCP. The difference in the gene expression of the receptors could be due to the different responsiveness to ethylene (and perhaps to 1-MCP) between the floral organs since the receptacles respond rapidly to ethylene by induction of *DcACS1* and *DcACO1* transcripts relative to petal tissues (Overbeek and Woltering, 1990 and Drory et al., 1993). Previous studies also showed that change in the receptor levels is associated with autocatalytic ethylene synthesis, which is mediated by the members of ACS family (Riov and Yang, 1982, Barry and Giovannoni, 2007 and Kevany et al., 2007).

DcETR1 and *DcERS1* rapidly decreased with the full flower opening and increased at onset of flower senescence. mRNA levels of *DcETR1* and *DcERS1* were up-regulated in the floral organs treated with ethylene while *DcERS2* transcript levels were unaffected by ethylene. Differences between ethylene receptors isoforms have also been observed in other species including *Arabidopsis* (Hua and Meyerowitz, 1998, Binder et al., 2004b and Chen et al., 2007), carnations (Shibuya et al., 2002), roses (Ma et al., 2006), tomato (Tieman et al., 2000 and Kevany et al., 2007), *Delphinium* (Tanase and Ichimura, 2006), and apple (Tatsuki et al., 2009). One potential interpretation is that these increases in gene-specific expression of *DcETR1* and *DcERS1* by ethylene may contribute to an increased sensitivity to ethylene, probably due to the decreased levels of the receptor proteins. However, we cannot rule out that receptor protein levels actually decrease, leading to increased ethylene sensitivity as reported in *Arabidopsis* and tomato (Chen et al., 2007 and Kevany et al., 2007). More refined analyses of receptor protein levels will be required to resolve this.

Overall, our results suggest that ethylene receptors are continually synthesized during later stages of floral development. While ethylene responsiveness is temporarily blocked by 1-MCP treatment, we predict that new receptors synthesized days after 1-MCP treatment can bind ethylene leading to the observed recovery of ethylene sensitivity in the flowers. In addition, we predict that the degradation of ethylene receptors that is stimulated by ethylene binding is prevented by successive treatments with 1-MCP prior to recovery of ethylene sensitivity. This study improves our understanding about the uses and effectiveness of 1-MCP.

5. CONCLUSIONS

Despite the initial block in ethylene perception with 1-MCP, carnation flowers regain ethylene sensitivity and resume endogenous ethylene biosynthesis over time. We were able to demonstrate that recovery of ethylene responsiveness is completely prevented by successive treatments of 1-MCP. In this study, the mRNA levels of *DcETR1* and *DcERS1*decreased during flower opening and increased at onset of flower senescence, suggesting that the receptors are generated during flower development and degraded in senescing tissues. We hypothesize that ethylene receptors are continually synthesized during flower development. Receptors that are newly generated after 1-MCP treatment can bind ethylene which would result in renewed sensitivity to ethylene. Since we observed that successive treatments with 1-MCP completely inhibited ethylene responses, we hypothesize that blocking ethylene binding to the both receptors that is newly synthesized or dissociated from ethylene results in continued inhibition of ethylene responses. By measuring expression in different tissue types, we were able to provide support for this hypothesis. Additional studies on receptor protein levels during similar treatments will be needed to further test this hypothesis.

ACKNOWLEDGEMENTS

This work was funded partially by AgroFresh and the Department of Horticulture, University of Wisconsin-Madison. We would like to thank Dr. Oh-Keun Kwon for technical comments on carnation growing and AgroFresh for supplying 1-MCP.

REFERENCES

1. Abeles, F.B., Morgan, P.W., Saltveit, J.M.E., 1992. Ethylene in Plant Biology. Academic Press, New York.

2. Barry, C.S., Giovannoni, J.J., 2007. Ethylene and fruit ripening. J. Plant Growth Regul. 26, 143–159.

3. Binder, B.M., Bleecker, A.B., 2003. A model for ethylene receptor function and 1- methylcyclopropene action. Acta Hortic. 628, 177–187.

4. Binder, B.M., Mortimore, L.A., Stepanova, A.N., Ecker, J.R., Bleecker, A.B., 2004a. Short-term growth responses to ethylene in Arabidopsis seedlings are EIN3/EIL1 independent. Plant Physiol. 136, 2921–2927.

5. Binder, B.M., O'Malley, R.C., Wang, W., Moore, J.M., Parks, B.M., Spalding, E.P., Bleecker, A.B., 2004b. Arabidopsis seedling growth response and recovery to ethylene. A kinetic analysis. Plant Physiol. 136, 2913–2920.

6. Blankenship, S.M., Dole, J.M., 2003. 1-methylcyclopropene: a review. Postharvest Biol. Technol. 28, 1–25.

7. Borochov, A., Woodson, W.R., 1989. Physiology and Biochemistry of Flower Petal Senescence, Horticultural Reviews. John Wiley & Sons, Inc., pp. 15–43.

8. Cameron, A., Reid, M.S., 2001. 1-MCP blocks ethylene-induced petal abscission of Pelargonium peltatum but the effect is transient. Postharvest Biol. Technol. 22, 169–177.

9. Chang, C., Kwok, S.F., Bleecker, A.B., Meyerowitz, E.M., 1993. Arabidopsis ethyleneresponse gene ETR1: similarity of productto two-component regulators. Science 262, 539–544.

10. Charng,Y.Y., Sun,C.W.,Yan, S.L.,Chou, S.J.,Chen,Y.R.,Yang, S.F., 1997. cDNAsequence of a putative ethylene receptor from carnation petals. Plant Physiol. 115, 863.

11. Chen, Y.F., Shakeel, S.N., Bowers, J., Zhao, X.C., Etheridge, N., Schaller, G.E., 2007. Ligand-induced degradation of the ethylene receptor ETR2 through a proteasome-dependent pathway in Arabidopsis. J. Biol. Chem. 282, 24752–24758.

12. Drory, A., Mayak, S., Woodson, W.R., 1993. Expression of ethylene biosynthetic pathway mRNAs is spatially regulated within carnation flower petals. J. Plant Physiol. 141, 663–667.

13. Ekman, J.H., Clayton, M., Biasi, W.V., Mitcham, E.J., 2004. Interactions between 1- MCP concentration, treatment interval and storage time for 'Bartlett' pears. Postharvest Biol. Technol. 31, 127–136.

14. Halevy, A.H., Mayak, S., 1981. Senescence and postharvest physiology of cut flowers. Part 2. Hortic. Rev. 3, 59–143.

15. Hall, A.E., Findell, J.L., Schaller, G.E., Sisler, E.C., Bleecker, A.B., 2000. Ethylene perception by the ERS1 protein in Arabidopsis. Plant Physiol. 123, 1449–1457.

16. Harada, T., Murakoshi, Y., Torii, Y., Tanase, K., Onozaki, T., Morita, S., Masumura, T., Satoh, S., 2011. Analysis of genomic DNA of DcACS1, a 1-aminocyclopropane-1- carboxylate synthase gene, expressed in senescing petals of carnation (Dianthus caryophyllus) and its orthologous genes in D. superbus var. longicalycinus. Plant Cell Rep. 30, 519–527.

17. Henskens, J.A., Rouwendal, G.J., ten Have, A., Woltering, E.J., 1994. Molecular cloning of two different ACC synthase PCR fragments in carnation flowers and organspecific expression of the corresponding genes. Plant Mol. Biol. 26, 453–458.

18. Hua, J., Meyerowitz, E.M., 1998. Ethylene responses are negatively regulated by a receptor gene family in Arabidopsis thaliana. Cell 94, 261–271.

19. Hua, J., Sakai, H., Nourizadeh, S., Chen, Q.G., Bleecker, A.B., Ecker, J.R., Meyerowitz, E.M., 1998. EIN4 and ERS2 are members of the putative ethylene receptor gene family in Arabidopsis. Plant Cell 10, 1321–1332.

20. Jiang, Y.M., Joyce, D.C., Macnish, A.J., 1999. Responses of banana fruit to treatment with 1-methylcyclopropene. Plant Growth Regul. 28, 77–82.

21. Jones, M.L., Woodson, W.R., 1999. Differential expression of three members of the 1-aminocyclopropane-1-carboxylate synthase gene family in carnation. Plant Physiol. 119, 755–764.

22. Jones, M.L., 2003. Ethylene biosynthetic genes are differentially regulated by ethylene and ACC in carnation styles. Plant Growth Regul. 40, 129–138.

23. Kende, H., 1993. Ethylene biosynthesis. Annu. Rev. Plant Phys. 44, 283–307.

24. Kevany, B.M., Tieman, D.M., Taylor, M.G., Dal Cin, V., Klee, H.J., 2007. Ethylene receptor degradation controls the timing of ripening in tomato fruit. Plant J. 51, 458–467.

25. Kieber, J.J., Rothenberg, M., Roman, G., Feldmann, K.A., Ecker, J.R., 1993. Ctr1, a negative regulator of the ethylene response pathway in arabidopsis, encodes a member of the Raf family of protein-kinases. Cell 72, 427–441.

26. Ma, N., Tan, H.,Xue, J.H., Li, Y.Q., Gao, J.P., 2006. Transcriptional regulation of ethylene receptor and CTR genes involved in ethylene-induced flower opening in cut rose (Rosa hybrida) cv. Samantha. J. Exp. Bot. 57, 2763–2773.

27. Macnish,A.J., Joyce, D.C., Iriving, D.E.,Wearing,A.H., 2004.Asimple sustained release device for the ethylene binding inhibitor 1-methylcyclopropene. Postharvest Biol. Technol. 32, 321–338.

28. Mor, Y., Halevy, A.H., Spiegelstein, H., Mayak, S., 1985. The site of 1-aminocyclopropane-1-carboxylic acid synthesis in senescing carnation petals. Physiol. Plant. 65, 196–202.

29. Nagata, M., Tanikawa, N., Onozaki, T., Mori, H., 2000. Ethylene receptor gene (ETR) homolog from carnation. J. Jpn Soc. Hortic. Sci. 1 (Suppl. 69), 407.

30. O'Malley, R.C., Rodriguez, F.I., Esch, J.J., Binder, B.M., O'Donnell, P., Klee, H.J., Bleecker, A.B., 2005. Ethylene-binding activity, gene expression levels, and receptor system output for ethylene receptor family members from Arabidopsis and tomato. Plant J. 41, 651–659.

31. Overbeek, J.H.M., Woltering, E.J., 1990. Synergistic effect of 1-aminocyclopropane- 1-carboxylic acid and ethylene during senescence of isolated carnation petals. Physiol. Plant. 79, 368–376.

32. Park, K.Y., Drory, A., Woodson, W.R., 1992. Molecular cloning of an 1-aminocyclopropane-1-carboxylate synthase from senescing carnation flower petals. Plant Mol. Biol. 18, 377–386. Reid, M.S., Wu, M.J., 1992. Ethylene and flower senescence. Plant Growth Regul. 11, 37–43.

33. Reid, M.S., Ç elikel, F.G., 2008. Use of 1-methylcyclopropene in ornamentals: carnations as a model system for understanding mode of action. HortScience 43, 95–98.

34. Riov, J., Yang, S.F., 1982. Autoinhibition of ethylene production in citrus peel disks – suppression of 1-aminocyclopropane-1-carboxylic acid synthesis. Plant Physiol. 69, 687–690.

35. Sakai, H., Hua, J., Chen, Q.G., Chang, C., Medrano, L.J., Bleecker, A.B., Meyerowitz, E.M., 1998. ETR2 is an ETR1-like gene involved in ethylene signaling in Arabidopsis. Proc. Natl. Acad. Sci. U.S.A. 95, 5812–5817.

36. Schaller, G.E., Bleecker,A.B., 1995. Ethylene-binding sites generatedinyeast expressing the Arabidopsis ETR1 gene. Science 270, 1809–1811.

37. Schaller, G.E., 2012. Ethylene and the regulation of plant development. BMC Biol. 10. Schotsmans, W.C., Prange, R.K., Binder, B.M., 2009. 1-Methylcyclopropene: mode of action and relevance in postharvest horticulture research. Hortic. Rev. 35, 263–313.

38. Serek, M., Sisler, E.C., Reid, M.S., 1994. Novel gaseous ethylene binding inhibitor prevents ethylene effects in potted flowering plants. J. Am. Soc. Hortic. Sci. 119, 1230–1233.

39. Serek, M., Sisler, E.C., Reid, M.S., 1995. Effects of 1-MCP on the vase life and ethylene response of cut flowers. Plant Growth Regul. 16, 93–97.

40. Shibuya, K., Satoh, S., Yoshioka, T., 1998. A cDNA endoding a putative ethylene receptor related to petal senescence in carnation (Dianthus caryophyllus L.) flowers. Plant Physiol. 116, 867.

41. Shibuya, K., Nagata, M., Tanikawa, N., Yoshioka, T., Hashiba, T., Satoh, S., 2002. Comparison of mRNA levels of three ethylene receptors in senescing flowers of carnation (Dianthus caryophyllus L.). J. Exp. Bot. 53, 399–406.

42. Sisler, E.C., Reid, M.S., Yang, S.F., 1986. Effect of antagonists of ethylene action on binding of ethylene in cut carnations. Plant Growth Regul. 4, 213–218.

43. Sisler, E.C., Blankenship, S.M., Guest, M., 1990. Competition of cyclooctenes and cyclooctadienes for ethylene binding and activity in plants. Plant Growth Regul. 9, 157–164.

44. Sisler, E.C., Dupille, E., Serek, M., 1996. Effect of 1-methylcyclopropene and methylenecyclopropane on ethylene binding and ethylene action on cut carnations. Plant Growth Regul. 18, 79–86.

45. Sisler, E.C., Serek, M., 2003. Compounds interacting with the ethylene receptor in plants. Plant Biol. 5, 473–480.

46. Tanase, K., Ichimura, K., 2006. Expression of ethylene receptors Dl-ERS1-3 and DlERS2. and ethylene response during flower senescence in Delphinium. J. Plant Physiol. 163, 1159–1166.

47. Tanase, K., Onozaki, T., Satoh, S., Shibata, M., Ichimura, K., 2008. Differential expression levels of ethylene biosynthetic pathway genes during senescence of long-lived carnation cultivars. Postharvest Biol. Technol. 47, 210–217.

48. Tatsuki, M., Hayama, H., Nakamura, Y., 2009. Apple ethylene receptor protein concentrations are affected by ethylene, and differ in cultivars that have different storage life. Planta 230, 407–417.

49. ten Have, A., Woltering, E.J., 1997. Ethylene biosynthetic genes are differentially expressed during carnation (Dianthus caryophyllus L.) flower senescence. Plant Mol. Biol. 34, 89–97.

50. Tieman, D.V., Taylor, M.G., Ciardi, J.A., Klee, H.J., 2000. The tomato ethylene receptors NR and LeETR4 are negative regulators of ethylene response and exhibit functional compensation within a multigene family. Proc. Natl. Acad. Sci. U.S.A. 97, 5663–5668.

51. Verlinden, S., Boatright, J., Woodson, W.R., 2002. Changes in ethylene responsiveness of senescence-related genes during carnation flower development. Physiol. Plant 116, 503–511.

52. Wang, W.Y., Hall, A.E., O'Malley, R., Bleecker, A.B., 2003. Canonical histidine kinase activity of the transmitter domain of the ETR1 ethylene receptor from Arabidopsis is not required for signal transmission. Proc. Natl. Acad. Sci. U.S.A. 100, 352–357.

53. Watkins, C.B., Nock, J.F., Whitaker, B.D., 2000. Responses of early, mid and late season apple cultivars to postharvest application of 1-methylcyclopropene (1- MCP) under air and controlled atmosphere storage conditions. Postharvest Biol. Technol. 19, 17–32.

54. Woodson, W.R., Lawton, K.A., 1988. Ethylene-induced gene-expression in carnation petals – relationship to autocatalytic ethylene production and senescence. Plant Physiol. 87, 498–503.

55. Woodson, W.R., Park, K.Y., Drory, A., Larsen, P.B., Wang, H., 1992. Expression of ethylene biosynthetic pathway transcripts in senescing carnation flowers. Plant Physiol. 99, 526–532.

56. Yang, S.F., Hoffman, N.E., 1984. Ethylene biosynthesis and its regulation in higher plants. Annu. Rev. Plant Phys. 35, 155–189.

Chapter 7

Effects of Long-term Nitrogen and Organic Fertilization on Antioxidants Content of Tomato Fruits

Enping Zhang*, **Yu Duan, Fulei Tan and Shuhong Zhang**

Department of Horticultural, Shenyang Agricultural University, 110866, China

ABSTRACT

The effects of different N level in combination with organic fertilizer on carotenoids, phenols and flavonoids contents in tomato fruits were studied. The N mixed with organic fertilizer treatment had higher the content of β-carotene, and AN2 achieved 34.20 μg/g. At red ripening stage, the content of lycopene of BN1 and BN2 were very close to, respectively 180.79 μg/g and 182.50 μg/g. The content of lutein at red ripening stage was nearly three times than that at turning stage. At red ripening stage, content of lutein ranged from 2.85 μg/g to 8.87 μg/g. In the all of phenolic acid, coffeic acid was the highest levels. The highest caffeic acid content (73.74 μg/g) was observed in the AN2 (double N and organic fertilizer), and only organic fertilizer (AN0) was no significant difference with BN1 (single N). Rutin content in tomato fruit had no difference in three N levels (N0, N1, N2). AN0 had the highest quercetin content in tomato fruits in all treatment, by 66.39 μg/ g.

KEYWORDS

Tomato; Carotenoids; Phenolic acid; Flavonoids; Fertilization

1. INTRODUCTION

Tomatoes are planted in worldwide because its fruits have many nutrients and high yields. As an important vegetable, they are an excellent source of **vitamin** C, vitamin E, folic acid, potassium, and antioxidants [1-3]. Since the early 1970s, antioxidants content in tomato fruits became a focus. Main antioxidants in tomato fruits include carotenoids, phenols and flavonoids. Immune system of body can reduce the incidence of cancers and other diseases by using carotenoids to adjust [4,5]. Meanwhile lycopene is providing protection against various cancers, angiocardiopathy, and other chronic diseases, probably due to its antioxidative properties [6]. Phenolic acids have stronger ability to resistant oxidation and damage of normal cells when it accumulates in body [7,8]. It can efficiently wipe out the singlet oxygen and peroxide free radicals of body. Flavonoids also can efficiently reduce radical which make the person aging to delay aging, because they have strong antioxidant activity.

Biosynthesis of lycopene and other carotenoids in plants comes from mevalonic acid and AcCoA. Plant phenol compounds are mainly through shikimic acid pathway and malonic acid pathway to synthetic. Flavonoids are the product of shikimic acid and more ketone ways, and biosynthesis of flavonoid is the basic skeleton of the three Malonyl CoA and 1 coumaric acyl coenzyme A (coumaroyl CoA). Those all are closely related to the nitrogen which is components of **enzymes** and plays an important role in metabolism and energy conversion [9-11]. So the antioxidant capability in tomato fruits is strongly affected by fertilization. But information on antioxidant contents of tomato affected by fertilization is scarce. Aldrich et al. and Anissa Riahi et al. focused on the total phenols content and lycopene content affected by different cultivars under fertilization. In addition, our research results showed that N combination with organic fertilizer can promote tomato yield and quality of tomato. However, in that study, antioxidant content of tomato fruits was not determined. Now the determination of antioxidant content in tomato fruits allows a real evaluation of nutritional quality of food rather than the analysis of basic sugar, protein, and basic content. In fact, these antioxidant contents contribute to the health benefits of body which people care about it [12-14].

In the last decade of the last century the excess of N becomes widespread in agriculture, namely as a consequence to pursuit of high yield and also led to the decrease of the quality of crops [15,16]. The objective of this study was to investigate response to different levels N fertilizers in combination with organic **fertilizers** on carotenoids, phenols and flavonoids contents in tomatoes fruits.

2. MATERIALS AND METHODS

2.1 Materials and treatments

Tomato (liaoyuanduoli) designed for field growing and test was examined during the year 2014 growing season (March-July). Tomato seedlings were grown under constant conditions of temperature 25°C/ 20°C (day/night), photoperiod 16 h light, 8h dark, illumination 250 μE/s/m. After 50 days the plants were transplanted to the plots in greenhouse.

The trial was done at the vegetable research station of Shenyang Agricultural University (Loc: 41°20′N, 123°38′E), China. The longterm fertilization experiment began in 1988, and the first 8.5 years proceeded in the open with a year of two stubbles and moved to an enclosed facility with a year of stubble in 1997. The fertility of foundation soil before the experiment was such that organic matter content was 24.3 g/kg, total N 1.164 g/kg, total P 1.374 g/kg, available N 86.41 g/kg, available P 70.8 g/kg, available K 56.14 g.kg and pH 6.75. This trial used to denote plots manure (AN0), manure +N fertilizer (AN1), manure +double N fertilizer (AN2) without fertilization (BN0), N fertilizer (BN1), double N fertilizer (BN2). A completely randomized design, replicated three times was selected with plot sizes of 1.5 m^2. To prevent fertilizers permeating into adjacent plots, each plot was built as a 0.8 m deep sealed cement pool. Manure was applied once before planting while N fertilizer was applied twice during the growing season. The experiment was finished after 4 month and half of plants growing.

Tomato fruits were hand harvested randomly from the rows and from the middle of each plant at three stages (turning, red, red ripening) and delivered quickly to the laboratory. Each tomato sample (whole fruit) washed, and then cut into small pieces and freeze-dried under an inert **nitrogen** atmosphere at -70°C. The freeze-dried samples were ground in a coffee grinder and screened through a size-20 standard sieve to obtain uniform particle size, then stored in deeper freezer until analyzed.

2.2 Chemical reagents

Standards (lycopene, β-carotene and lutein), objects, and BHT were purchased from Shanghai Yuanye technological company. Phenolic acids and standards (caffeic acid, ferulic acid, cinnamon acid, chlorogenic acid, benzoic acid, anthocyanin, rutin, quercetin, luteolin, kaempferol) were purchased from Sigma. All of HPLC-grade solvents were purchased from Fisher Chemicals.

2.3 Analytical procedures

Determination of carotenoid: 0.2 g homogenate was used for carotenoid extraction. The mixture of acetone: petroleum ether (2:1) was used for carotenoid extraction, adding 10 μg internal standard substance. The vials were then placed in a sonicator bath at 30°C for 30 min. The mixture was centrifuged and the supernatant was transferred into a 10 mL tube. After vacuum freeze-dried of the supernatant, using 1.0 mL ethyl acetate dissolved and then filter the 0.45 μm micro porous filter membrane. The filter was used for HPLC analysis [17].

Column: Nova-pak C18, 4 μm, 3.9 mm × 150 mm; temperature: 30°C, mobile phase: acetonitrile: trichloromethane (92:8), flow rate:1.0 ml per minute, sample loop 20 μl, spectrophotometric detector λ=450 nm. Standard of lycopene, β-carotene and lutein was used for peak identification and quantification.

Determination of phenol: For each extraction, approximately 0.2 g of ground freeze-dried tomato sample was placed in a 20 ml tube with 5 mL of the solvent mixture ethanol:H_2O (50:50,%v/v). The vials were then placed in a sonicator bath at 60°C for 20 min.The mixture was centrifuged and the supernant was transferred into a 10 mL tube. The residue was resuspended in 5 mL of mixture (above) and mixed, then sonicated for an additional 20 min followed by centrifugantion. The supernant was combined with the initial extract and the volume of combined wupernant was made up to 10 mL with the extraction solvent and waited for HPLC [18].

Table 1. Treatment of experiment.

Treatment	Horse manure(g/plot)	Urea (g/plot)
AN0	11.25	0
AN1	11.25	97.8
AN2	11.25	195.6
BN0	0	0
BN1	0	97.8
BN2	0	195.6

Phenolic acids were analyzed on a HPLC. Separation of phenolic acids was achieved on a reversed phase C18 Luna column (Phenomenex, 3.9 mm × 150 mm; particle size 4 μm). The column was thermostatically controlled at 30°C and the flow rate was set to 1.0 mL/ min. The mobile phase consisted of two solvents: 3% (ethylic acid (A) and methanol (B). Two wavelengths (280 nm and 320 nm) were used to detect the eluent composition. All phenolic acids were quantified with external standards by using HPLC analysis.

2.4 Determination of flavonoid

For each extraction, approximately 0.2 g of ground freeze-dried tomato sample was placed in a 20 ml tube with 5 mL of the solvent mixture ethanol: H_2O (50:50,%v/v). The vials were then placed in a sonicator bath at 60° for 20 min.The mixture was centrifuged and the supernant was transferred into a 10 mL tube. The residue was resuspended in 5 mL of mixture (above) and mixed, then sonicated for an additional 20 min followed by centrifugation. The supernatant was combined with the initial extract and the volume of combined supernatant was made up to 10 mL with the extraction solvent and waited for HPLC.

Column: Nova-pak C18, 4 μm, 3.9 mm × 150 mm; temperature: 35, mobile phase: 0.5% phosphoric acid:methanol:(1:1), flow rate:1.0 ml per minute, sample loop 10 μl, spectrophotometric detector λ=360 nm.

2.5 Statistical analysis

Effect of fertilizer treatments on **antioxidant** in tomato fruits were assessed by analysis of variance (ANOVA) and comparison among means was determined according to Tukey's test.

3. RESULTS

3.1 Carotenoids content

Influence of fertilization on β-carotene content is given in Figure 1. Compared to the single N treatment, the N mixed with manure treatment had higher the content of β-carotene, among of those, AN2 achieved 34.20 μg/g. Followed the tomato fruits mature, the content of β-carotene was dramatically increased. Different level N showed different content of β-carotene. In all, the more N number, the higher content of β-carotene.

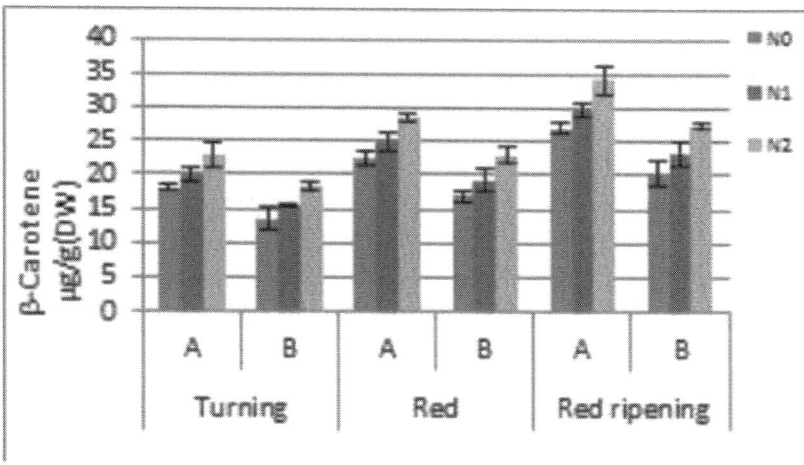

Figure1. Carotenoids content in tomato fruits under di erent N levels mixed with organic fertilizer.

Influence of fertilization on lycopene content is given in Figure 2. Followed the tomato fruits mature, the content of lycopene was increased. There was no significantly different content of lycopene between the red and red ripening stage of fruits, but they were all higher than the turning stage. N with manure increased the content of lycopene compare with single N, and the more N number, the higher content of lycopene. But at red ripening stage, the content of lycopene of BN1 and BN2 were very close to, respectively 180.79 μg/g and 182.50 μg/g that was to say over N level had nothing to do with lycopene content. But N mixed with organic, higher N level increased lycopene content, so AN2 had the highest content of lycopene, 238.42 μg/g.

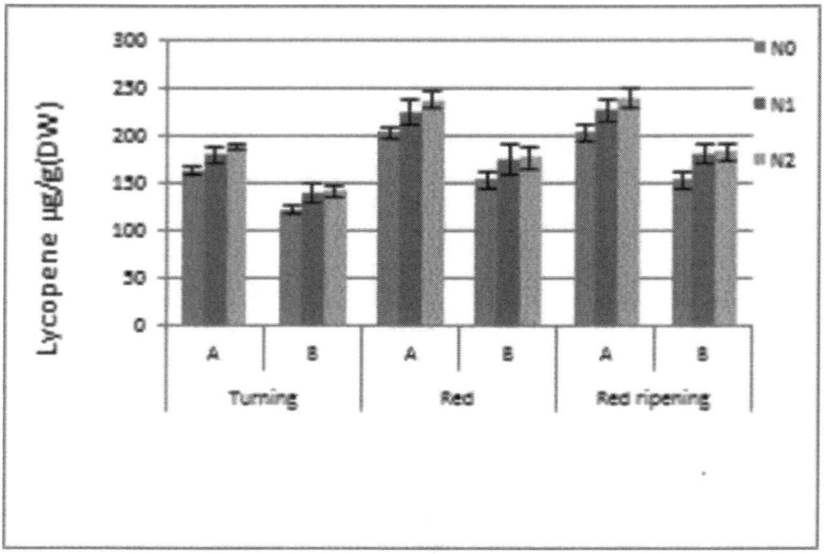

Figure 2. Lycopene content in tomato fruits under diﬀerent N levels mixed with organic fertilizer.

Influence of fertilization on content of lutein is given in Figure 3. Followed the tomato fruits mature, the content of lycopene was increased. The content of lutein at red ripening stage was nearly three times than that at turning stage. N fertilizer and mixed with manure could increase the content of lutein compared to only N fertilizer. At red ripening stage, content of lutein ranged from 2.85 μg/g to 8.87 μg/g, AN0 to AN2. The content of lutein of only manure (BN0) was higher than that of sing N **treatment** (AN1), and had on difference with the double N fertilizer (AN2).

3.2 Phenolic acid content

Phenolic acids and total phenol contents of tomato fruits grown under the different fertilizer treatments are shown in Figures 4 and 5, respectively.

The contents of caffeic acid, cinnamon acid, ferulic acid, chlorogenic acid, benzoic acid and anthocyanin in tomato varied depending on the different N level and mix with organic fertilizer. In general, all these phenolic compounds increased after N mixed with organic fertilizer, and improved with N level increased. In the all of phenol acid, coffeic acid is the highest levels. The highest caffeic acid content (73.74 μg/g) was observed in the AN2 (double N and manure), and only organic fertilizer AN0 was no significant difference with BN1 (single N). Benzoic acid and anthocyanin were the lowest levels in phenolic acid.

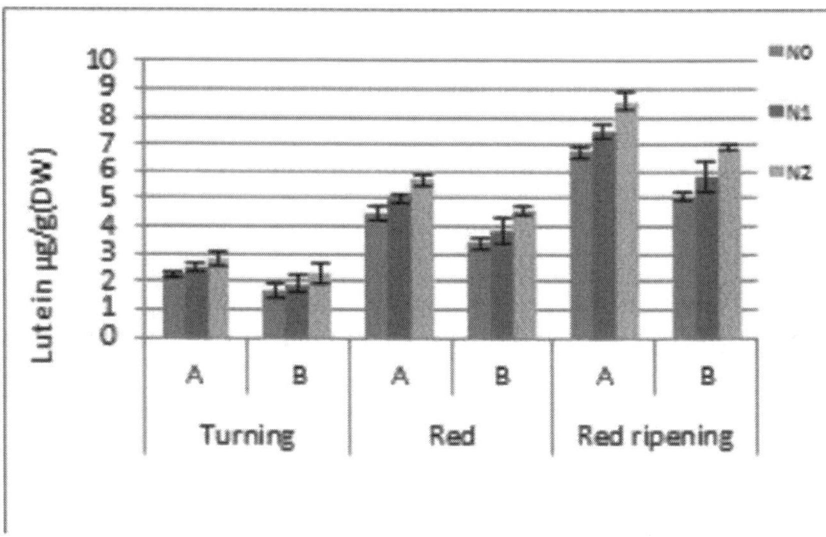

Figure 3. Lutein content in tomato fruits under different N levels mixed with organic fertilizer.

Total phenol contents of AN1 and AN2 had no difference, but higher than that BN2, 250.7 μg/g, 257.12 μg/g respe ctively.

Flavonoids detected in tomato including rutin, quercetin, and Luteolin, kaempferol, among them the highest content of rutin, the lowest content of kaempferol. Different nitrogen levels had different effects on content of flavonoids. Organic with N fertilizer increased different flavonoids contents, higher than only N ferlilizer. So, organic fertilizer had significant effects on flavonoid content. Rutin content of fruit of tomato had on difference in three N levels (N0, N1, N2). AN0 had the highest quercetin content of tomato fruits in all treatment, by 66.39 µg/g, but the BN0 had the lowest content, only 53.65 µg/g. Under only N fertilizer, with the increase level N, the content of quercetin, luteolin, and kaempferol increased as well.

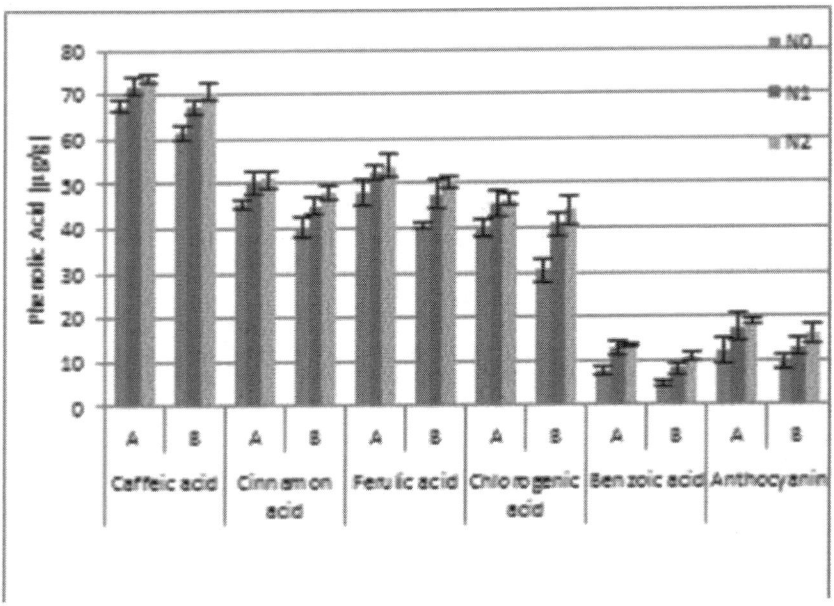

Figure 4. Phenolic acids content in tomato fruits under different N levels mixed with organic fertilizer.

4. DISCUSSION

In the different stage of fruit, β-carotene content in tomato fruit showed a rising trend, and at turning stage β-carotene content in the fruit was low just because its coloring just began to form. But β- carotene content increased rapidly at red stage, mainly due to β- carotene fully formed [19,20]. The results of this study showed that fully mature tomatoes under different N levels forms the accumulation of carotenoids, its main content was lycopene, about 80% of the total carotenoid content, followed by β-carotene content of only about 20% of the total carotenoid content, lutein content at least. But this study we used only one tomato varieties, carotenoid content in different colors, different varieties may vary, so the mechanism needs further to study. The results showed that from turning to red ripening, the development of the lycopene content in tomato fruit changed significantly, and that synthetic lycopene, are the fruits of lycopene synthesis in critical period, this is consistent with most of the scholars ' research [21,22]. Letein content under different N levels in the whole growth period was on the rise, and N mixed with organic fertilizer had greater effects than the inorganic fertilizer treatment.

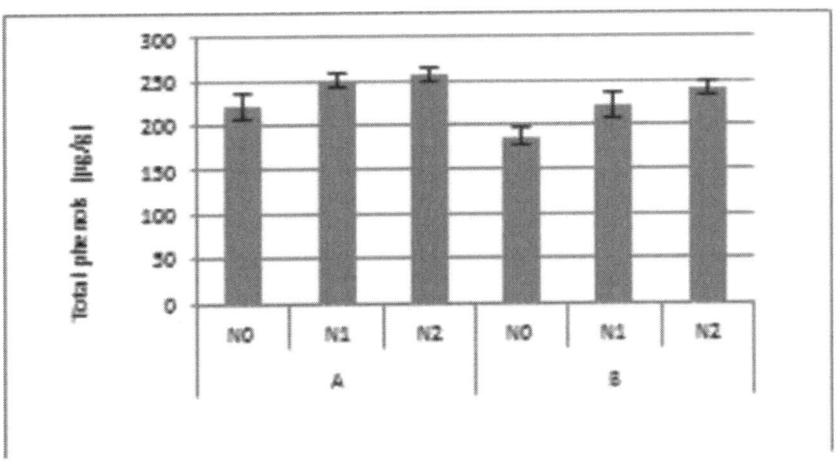

Figure 5. Total phenos content in tomato fruits under di̵erent N levels mixed with organic fertilizer

Plants rich in many types of phenolic compounds and its structure are different too [23]. Studies have shown that with the increase of fruit maturity increased phenolic compounds in fruits; this is consistent with the conclusions of this experiment [24]. Studies have shown that different growth conditions and different temperature light will lead to the different content of phenols and different constitute [25]. External conditions on the content of phenolic compounds in plants also have an impact [26]. This experiment selected common phenolic compounds in tomato fruit were detected and found phenolic compounds include caffeic acid, chlorogenic acid, ferulic acid and cinnamic acid, benzoic acid and anthocyanin. As the test for liquid chromatography and found the presence of unknown peak, so tomato fruit may also have other types of phenolic compounds in existence, their specific situations to be in later generations studied. Different from the results of this test, this may be associated with years of fertilization. Effects of different nitrogen levels on phenolic compounds in fruit accumulation and synthesis, specifically nitrogen can promote the synthesis of phenolic compounds. Applying organic fertilizer, most phenols content showed a facilitating role, but little effect on the acid and anthocyanin contents. Different levels of nitrogen fertilizer and organic fertilizer can increase the content of phenols, and twice times the nitrogen fertilizer and organic fertilizer under the condition of phenolic acids content in tomato fruits reaches its maximum value.

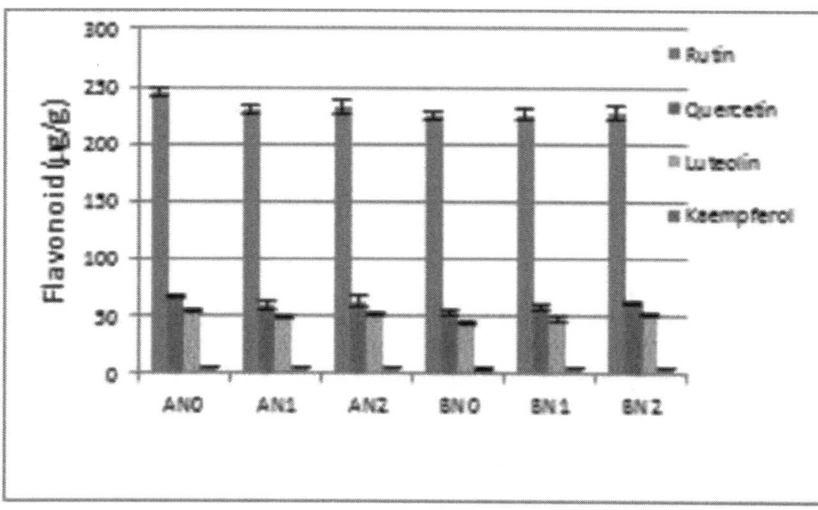

Figure 6. Flavonoid content in tomato fruits under di erent N levels mixed with organic fertilizer.

Flavonoids of synthesis and accumulation are a complex physiological and biochemical processes, this process requires many different genes to control [27]. Process of synthesis and accumulation of flavonoids, enzymes play a crucial role [28]. These enzymes include: flavanone reductase, isoflavone syntheses, chalcone reductase, etc., The activity of these enzymes will affect total flavonoids from fruits of synthesis and accumulation. We have previously found that the flavonoid content of tomato varies duting p production season dut to both abiotic (e.g., light) and agronomic (e.g., fertilization) factors [29]. Different minerals can change solutions of phenylalanine ammonialyase activity. Studies have shown that with the increase of fruit maturity flavonoids in fruits increased, consistent with the conclusions of this experiment. Studies confirm that a moderate reduction of application of inorganic fertilizer can increase plant flavonoids accumulation and synthesis. In N-fertilizer treatment, the flavonoids content in tomato fruits under N2 treatment had higher than that of N1 level, specifically N2>N1>N0. Mixture of organic fertilizer had significant effects on flavonoid content; diagrams can be found that combined application organic fertilizer can increase the content of flavonoids in fruit.

5. CONCLUSION

This current study highlights the difference in antioxidant content including carotenoids, phenols, and flavonoids contents of tomato grown in different N level and different fertilizer combinations. The results indicated higher level N had a higher level of carotenoids, in particular the β-carotene. Organic fertilizer can promote the phenols and flavonoids content, especially mixed with N. So we can mention that we should use organic fertilizer mixed with chemical fertilizer other than only chemical fertilizer, which not only improve tomato quality, but also build soil fertility.

REFERENCES

1. Ninfali P, Mea G, Giorgini S, Rocchi M, Bacchiocca M (2005) Antioxidant capacity of vegetables, spices and dressings relevant to nutrition. British J of Nutrition 93, 257–266.

2. George B, Kaur C, Khurdiya DS, Kapoor HC (2004) Antioxidants in tomato (Lycopersicon exculentum) as a function of genotype. Food Chem 84, 45-51.

3. Mitchell AE, Hong YJ, Koh E, Barrett KM , Bryant DE, et al. (2007) Ten-year comparison of the influence of organic and conventional crop management practices on the content of flavonoids in tomatoes. J Agric FoodChem 55,6:154-6159.

4. Le-Gall G, Du Pont MS, Mellon FA, Davis AL, Collins GJ, et al.(2003) Characterization and content of flavonoid glycosides in genetically modified tomato (Lycopersicon esculentum) fruit. J Agric Food Chem 51: 2438–2446.

5. Llahy R, Chafik H, Marcello L, Tlili L, Giuseppe D (2010) Antioxidant activity and bioactive compound changes during fruit ripening of high-lycopene tomato cultivars. J of Food Compostion and Analysis 24: 588-595.

6. Giovannucci E (1999) Tomatoes, tomato-based products, lycopene and cancer. Review of the epidemiological literature. J of National Cancer Institute 91: 317–331.

7. Kähkönen M, Hopia A, Vuorela H, Rauha J, Pihlaja K, et al.(1999) Antioxidant activity of plant extracts containing phenolic compounds. J of Agricultural and Food Chemistry 47: 3954–3962.

8. Rao AV, Agarwal S (2000) Role of antioxidant lycopene in cancer and heart disease. J Am Coll Nutr 19: 563-569.

9. Brunhold C (1993) Regulatory interactions between sulfate and nitrate assimilation. In:De. KokL J, Stulen I, Rennenberg H, Brunold C, Rauser WE (edition): Suplhur Nutrition andAssimilation in Higher Plants. SPB Academic Publishing, The Hague: 61-75.

10. Dumas Y, Dadomo M, DiLucca G, Grolier P (2003) Effects of environmental factors and agricultural techniques on antioxidant content of tomatoes. J Sci Food Agric 83: 369–382.

11. Yahia EM, Barrera A (2010) Antioxidant capacity and correlation with phenolic compounds and carotenoids in 40 horticultural commodities. Acta Horticulturae 877: 1215–1220.

12. Aldrich H, Salandanan K, Kendall P, Bunning M, Stonaker F, et al.(2010) Cultivar choice provides options for local organic and conventional production of more nutritious tomatoes. J Sci Food Agric 90: 2548-2555.

13. Anissa Riahi, Chafik H dider (2013) Bioactive compounds and antioxidant activity of organically grown tomato (Solanum lycopersicum L.) cultivars as affected by fertilization. Scientia Horticulturae 151: 90-96.

14. Pellegrini Nolombi B, Salvatore S, Brenna OV, Galaverna K, DelRio D et al.(2007) Evaluation of antioxidant capacity of some fruit and vegetable foods. Efficiency of extraction of a sequence solvents. J Sci Food Agric 87:103-111.

15. Tei F, Benincasa P, Guiducci M (2002) Critical nitrogen concentration in processing tomato. Eur J Agron 18: 45–55.

16. Seliga JP, Shattuck VI (1995) Crop rotation affects the yield and nitrogen fertilization response in processing tomatoes. Sci Hortic 64:159–166.

17. Ma YJ, Zhou BL, Yu Y, Fu YW (2010) Dynamic changes of carotenoids contents during ripening in tomato fruit and regulatory effect CEPA. Acta Horticulturae107-112.

18. Zhang EP, Yu XZ, Zhang SH (2012) Effect of Long-term Fertilization on the Material Accumulation and Distribution and Photosynthetic Characteristics of Tomatoes. The 3rd conference on horticulture science and technology 77-81.

19. Poiroux-Gonord F, Fanciullino AL, Bert L, Urban L (2012) Effect of fruit load on maturity and carotenoid content of clementine (Citrus clementina Hort. exTan.) fruits. J Sci Food Agric 92: 2076–2083.

20. Mercadante AZ, Rodriguez-amaya DB (1998) Effects of ripening, cultivar differences and processing on the carotenoid composition of mango. J Food Chem 8561:128–130.

21. Qu RF, Liang Y, Gong ZH, Yang YZ (2006) Variance of lycopene contents in different tomato varieties with fruit development. Acta Agri Boreali-Sin.15:121-123.

22. Fraser PD, Truesdale MR, Bird CR, Schuch W, Bramley PM (1994) Camtenoid biosynmesis during tomato fruit development (Evidence for Tissue-Specific Gene Expression) Plant Physiol 105: 405-413.

23. Hunt GM, Baker EA (1980) Phenolic constituents of tomato fruit cuticles. Phytochemistry 19: 1415–1419.

24. Tomás-Barberán F, Espín J (2001) Phenolic compounds and related enzymes as determinants of quality in fruits and vegetables. J of the Science of Food and Agriculture 81 853–876.

25. Chang C, Lin H, Chang C, Liu Y (2006) Comparisons on the antioxidant properties of fresh, freeze-dried and hot-air-dried tomatoes. Journal of Food Engineering 77: 478–485.

26. Chong CH, Law CL, Figiel A, Wojdyło A, Oziembłowski M (2013) Colour, phenolic content and antioxidant capacity of some fruits dehydrated by a combination of different methods. Food Chemistry 141: 3889-3896.

27. Cuyckens F, Claeys M (2004) Mass spectrometry in the structural analysis of flavonoids. J Mass Spectrom 39: 1–15.

28. Considine M, Daley D, Whelan J (2001) The expression of alternative oxidase and uncoupling protein during fruit ripening in mango. Plant Physiol 126: 1619–1629.

29. Slimestad R, Verheul MJ (2005) Seasonal variations in the level of plant constituents in greenhouse production of cherry tomatoes. J Agric Food Chem 53: 3114-3119.

Chapter 8

Increasing Fresh Edamame Bean Supply through Season Extension Techniques

Nolen S[1], Zhang B[2] and Kering MK[1*]

[1]Agricultural Research Station, Virginia State University, Petersburg, Virginia, USA

[2]Virginia Polytechnic Institute and State University, Blacksburg, USA

ABSTRACT

Soybean [Glycine max (L.) Merr.], a legume native to East Asia contains up to 40% protein and 20% oil. Edamame, a popular vegetable soybean in East Asia, especially China and Japan, harvested at reproductive stages six and seven (R6 or R7) is gaining popularity in the US. Increased awareness of its nutritional quality through promotional campaigns and changing population demographics in the US have led to recent raise in US market demand for edamame. To meet the increasing market demand, frozen edamame from China and Taiwan has been imported. However, the quality of such imported product quickly deteriorates under frozen condition. The objective of this study was to determine whether off-season production systems and staggered planting of different maturity groups (MG) edamame cultivars during the planting season can extend the harvesting window for fresh beans. Four released edamame cultivars of different maturity groups (MG) were used: Gardensoy31 (MG III), Gardensoy41 (MG IV), Mooncake (MG V) and Randolph (MG VI). Total pod yield, marketable pod yield and seed quality traits including protein, oil and sucrose content of each cultivar were determined. The results indicated that early- and mid-spring planting of all MG varieties in the high tunnels allows for pod harvest starting in early July. Planting early in plasticcovered field followed by conventional planting in late-spring allowed harvest in mid-summer through early fall. While total and marketable pod yield differed among cultivars and production systems, seeds had comparable oil content (158 g kg-1) and protein and sucrose content range of 370-422 g kg-1 and 33-73 g kg-1 on dry matter, respectively. Use of season extension production techniques and soybean of appropriate MG increases harvest window from two weeks to several months.

KEYWORDS

Soybean; Protein; Soil

1. INTRODUCTION

Japanese vegetable soybean (edamame) is harvested at reproductive (R) growth stage six to seven (R6 to R7) when the seeds are still green [1]. At this developmental stage, the seed contains high protein, monounsaturated fatty acids, minerals elements, as well as vitamins like B1 and B2 [2,3]. Edamame, a historically Japanese delicacy has been rapidly incorporated in to American diets due to changing ethnic diversity among US population demographics and increased people's awareness of edamame nutritional qualities. Edamame is currently found in grocery stores and farmers' markets with its demand in the U.S. being estimated at 14,877 tons annually [4,5]. Edamame production is a promising opportunity to meeting the specialty crop market demands. While over 70% of the edamame sold in supermarkets across the U.S is imported from China and Taiwan, it is mainly marketed frozen and this frozen storage conditions lowers its quality drastically. Commercial frozen edamame may contain harmful bacteria such as Escherichia coli and Listeria monocytogenes [6]. In addition, consumers raise the concern of edamame source of safety and start to seek locally grown fresh edamame [7].

Lack of long-term, local supply has become a barrier for successfully marketing of edamame in the U.S because it has a very narrow field harvest window of only a few days. There is very limited information available on extending edamame harvest window though season extension production techniques in Virginia. However, growth studies under plastic film house culture and plastic film tunnel on two Japanese edamame cultivars has been done elsewhere [8]. By planting from February 15 to September 25, they were able to harvest crop from May 25 to December 3. In the US, an intensive research was conducted to extend edamame harvest window in 2003 and 2004 at University of Kentucky [9]. Four Gardensoy series edamame cultivars from MG I to IV were planted in greenhouses and transplanted into the field either covered by plastic film or with no plastic protection on April 1st, 15th, and 29th in Kentucky [9]. The crop was harvested for fresh pods from June 17th (GardenSoy 11, MG I) to August 10th (GardenSoy 41, MG IV). In the study, GardenSoy 11 had lower marketable yield than other cultivars at all transplanting dates, and indication showed that early production by transplanted MG I cultivars may not be successful in Kentucky. Because no research has been done in Mid-Atlantic region to extend fresh edamame harvest window, growers are looking for superior edamame cultivars with high stability and adaptability. There is a possibility for increased on-farm income from edamame relatedbusinesses through off-season supply. To be able to supply fresh edamame for an extended period of time, there is a need for season extension system development.

2. MATERIALS AND METHODS

The experiment was carried out for two growing season (2013-2014) at Randolph Research and Demonstration Farm, Virginia State University. The study was laid out as a randomized complete block design with three replications. Four edamame cultivars of different maturity groups (MG) where used: GardenSoy31 (MG III), GardenSoy41 (MG IV), Moon Cake (MG V) and Randolph (MG VI. Edamame planting in early- and mid-spring was done in a high tunnel (a polyethylene-covered semi-circular structure). Planting date varied because of unforeseen weather conditions, however, it was categorized into early-spring (Before April 21th), mid-spring (April 21th-May 21th) and late spring (May 21th-June 21th), and late summer (Aug 21th-Sept 21th). During early and mid-spring planting in the high tunnel, the four edamame cultivars; Gardensoy31, Gardensoy41 and Moon cake Randolph were grown in seed beds covered with a black plastic.

During mid-spring planting in the field, edamame was grown in both plastic-covered and non-covered seed beds. Another planting in the field with or without plastic-cover was carried out in late spring. Also in late August-early September, crop was established in the high tunnel to protected crop from cold temperatures later in the fall.

During harvest, all plants were harvested and pods removed. Because cultivars of different MG were planted on the same time, harvesting was sequential. Within a production system, the first to mature were those belonging to the low MG category. Total weight of harvested pods was determined and then sorted into pods with 1, 2, 3 or 4 seeds. The weight of pods in the individual categories was then obtained. Total weight of marketable pods was obtained as weight of all pods with two or more seeds pod-1. A representative sample of the marketable pods was obtained, weighed and shelled. Number of seeds was obtained, and its fresh weight was determined. About 100 gram of the shelled seeds were obtained and stored frozen to await freeze drying. The freeze dried material was ground later and analyzed to determine oil, protein, and sucrose content. Protein content was determined using the combustion method (AOAC - Official Method 990.03) with a Vario MAX CN (Elementar Americas, Inc., Mt. Laurel, NJ, USA) as described by Association of Official Analytical Chemists [10]. Protein factor of 6.25 is used for calculation. Oil was extracted from dried and grounded samples using petroleum ether in an ANKOM XT15 Extractor, Method 2, 01-30-09 (ANKOM Technology, Macedon NJ, USA) as described by AOCS [11] Official Procedure Am 5-04. Sugars were extracted from ground sample (1 g) and analyzed by HPLC following the methods optimized by Johansen et al. [12]. Sugars in the extracts were identified by comparing their retention times with standard sugars. For quantification, trehalose was used as internal standard and the sugar concentration was expressed as g/100 g dry basis. Statistical analyses were carried out using SAS procedures.

3. RESULTS AND DISCUSSION

3.1 Plant growth and harvest window extension

Plants grown in high tunnel on plastic covered seed beds in both early- and mid-spring showed good growth. For mid-spring planting, while crop planted in the field on plastic-covered beds showed faster growth than those without plastic cover (Figure 1) and both were slower than that in the high tunnel. Similar use of high tunnels and other season extension techniques have been used in other horticultural crops [13,14] and its beneficial effects is a result of microclimate modification allowing crops to grow outside of their normal time schedule.

Figure 1. Growth in field plants for mid-spring planting with-(a) and without (b) plastic covered seed bed.

For crops grown in early spring, pod harvest was achieved as early as the beginning of July. In early July, Gardensoy31, a MG III vegetable soybean cultivar was harvested and Gardensoy41, MG IV was harvested about two weeks later. At the time of harvest of Gardensoy41 from early-spring planting, Gardensoy31 in the high tunnel from midspring planting was ready for harvest. Using the production systems (High tunnel, plastic covered seed- beds in the field, and conventional soybean production) and planting edamame cultivars of different maturity groups simultaneously extended the harvest window from a few days common for conventionally produced crop to several months (Table 1). Within a cultivar, harvesting occurred two weeks to a month between first and second crop in the high tunnel. For the same cultivar in the field, a crop on plastic-covered beds was harvested earlier than those without plastic, because it grew faster and matured earlier. Because of differences in maturity time attributed to planting dates, cultivar's MG, and use of multiple production systems, pod harvest occurred from early July and continued through end of September/

early-October (Table 1). While differences in MG was responsible for harvest time differences among cultivars planted at the same time under a given production system, presence or absence of plastic cover was responsible for differences within a cultivar. In general, Gardensoy31 (MG III) and Gardensoy41 (MG IV) were harvested earlier than Mooncake (MG V) and Randolph (MG VI) when planted at the same time under similar production system. However, extremely cold temperature and animal pest destroyed edamame grown in late summer for two successive years. While this crop was not included in the result, there was a potential to extend harvest into early November.

3.2 Pod yield and yield-related traits

For a specific planting period and for a specific production system (high tunnel, field with or without plastic covered seed-bed), total harvested pods differed with edamame cultivars. For early spring planting in the high tunnel, total pod yield varied from 4300 kg ha-1 in Gardensoy31 to over 10,000 kg ha-1 for Randolph (Table 2). Slightly higher yields were obtained during the second planting in the high tunnel in mid- spring. The increase may be attributed to increased soil and air temperatures that may have allowed for better growth conditions since this was planted 2-3 weeks later in the spring. Yield in a crop established in August and September in the high tunnel are not included due to loss of crops from destruction by Marmota monax (ground hog), an animal pest.

In mid-spring, though not significant, marketable yield from crop established in high tunnels or in the field with plastic covered seed-bed were relatively higher than that from field without plastic cover (Table 3). Overall, Randolph produced relatively high total pod yield than other cultivars within a given production system. In open field for mid-spring planting, Randolph yield range from 14000 to 23000 kg ha-1 fresh pods (Table 3). In late spring under conventional planting, Randolph produced the largest (P=0.05) total pod yield of 15850 kg ha-1. The other three varieties produced similar total pod yields averaged at 9900 kg ha-1 (Figure 2). The marketable yield for late spring planting ranged from 63 to 77 % of total pod yield depending on variety (Figure 2). The yields for same varieties in this study were lower than those obtained by others in Mississippi [5].While marketable yield for mid-spring planting range from a low of 4120 kg ha-1 in Gardensoy31 to 6545 kg ha-1 in Randolph under the high tunnel, it ranged from 2092 kg ha-1 in Gardensoy31 to 10,814 kg ha-1 in Randolph in field grown crop (Table 3 and Figure 2). These values for marketable yield is comparable to those of other vegetable soybean obtained other studies [15,16]. The difference in yield between cultivars within a production system could be attributed to differences in cultivar yield potentials. In field grown crop, plastic covered-bed increased yield in all cultivars. Under all the systems, marketable pods (≥ 2 seeds pod-1) were between 40-75% of the total yield. The number of pods with at least two beans was proportionally higher than the other class categories and could explain the high proportion of marketable pod yield. For all varieties, there was a strong correlation between total and marketable yield ($r>0.75$) (data not shown). There was also strong correlation ($r>0.75$) between total pod yield or marketable pod yield and number of pods with one, two, or three seeds.

This would be expected since total yield include all pod while marketable consist of those with ≥ two seeds pod-1and makes sense that as number of these pods in harvest increase, total or marketable yield also increases. Similar correlation has been shown before in other vegetable soybean varieties [17].

The oil content of fresh seed harvested from different production systems and varieties were similar (P-=0.05) and averaged 158 g kg-1 (Table 4). The sucrose content was affected by the interaction of the production system and variety ($P<0.05$).

In general, sucrose content was higher for plants seeded later in the field averaging 57 g kg-1 across varieties. Across varieties, sucrose content of seed from plants in the high tunnel averaged 43 g kg-1. The protein and oil content of seed in this study were comparable to those reported elsewhere [16,18,19]. While sucrose content were comparable to those reported before [20,21], they are higher than those of vegetable varieties reported in other studies [22]. Seed protein content was affected by interaction of the production system and variety ($P<0.05$). While there was observed a general increase in seed protein content with time from mid-spring to late-spring in Gardensoy31 and Gardensoy41, there was a reduction in content for Randolph (Table 4). Across production systems, Randolph seed protein content averaged at 396 g kg-1 was larger ($P=0.05$) compared to other varieties.

Table 1. Planting time and harvest period for the dLjerent edamame varieties produced under given production systems. Key ____________________Approximate period within which Gadensoy31 and Gardensoy41 where harvested in that order. -------------------- Approximate period within which Mooncake and Randolph where harvested in that order.

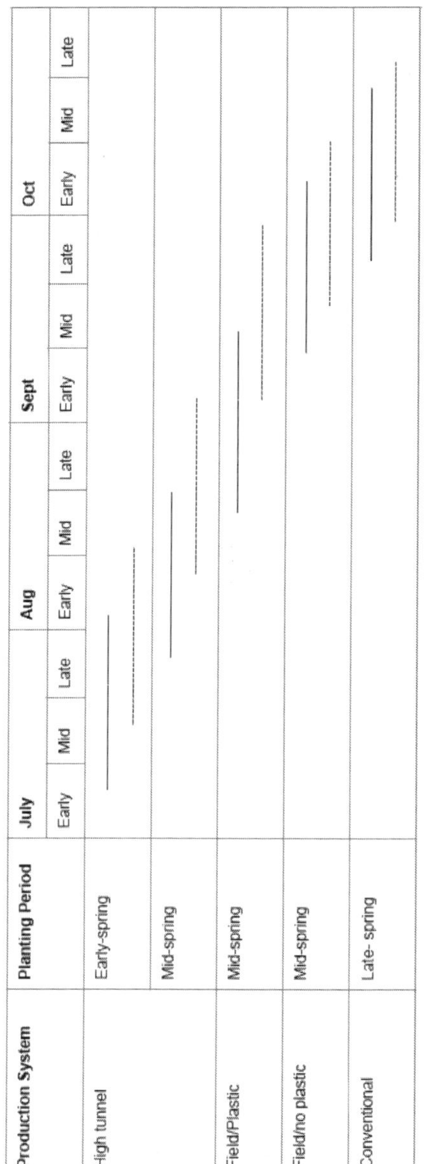

Table 2. Comparison of total and marketable pod yield for edamame cultivars grown under the high tunnel at different planting dates.

Planting period	**Cultivar**	**Harvested pods (kg ha^{-1})**	
		Total yield	Marketable yield
Early-spring	Gardensoy31	4373b	2487b
	Gardensoy41	7551ab	3915ab
	Mooncake	9838ab	5020ab
	Randolph	12957a	7506a
Mid-spring	Gardensoy31	6117ab	4120ab
	Gardensoy41	11340ab	6348ab
	Mooncake	10505ab	6053ab
	Randolph	9432ab	6545ab

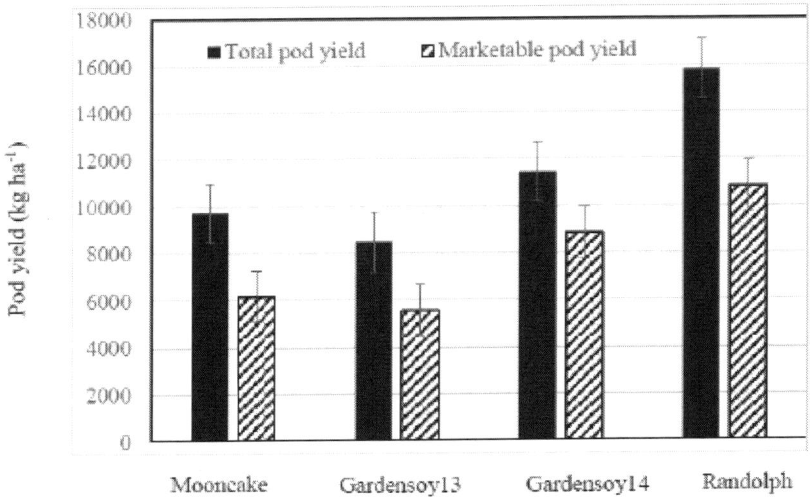

Figure 2. Total and marketable pod yield for crop under conventional planting in late spring. Values are means ± SE.

Table 3. Comparison of total and marketable pod yield in edamame cultivars planted in mid- spring using different production systems.

Production system	**Cultivar**	**Harvested pods (kg ha^{-1})**	
		Total yield	Marketable yield
High tunnel	Gardensoy31	6117c	4120cd
	Gardensoy41	11340bc	6348bcd
	Mooncake	10505bc	6053bcd
	Randolph	9432bc	6545bcd

Production system	Time of Planting	Variety				
		Gardensoy31	Gardensoy41	Mooncake	Randolph	Mean
		____________(g kg^{-1})____________				
		----------------Protein----------------				
High tunnel	Early-spring	370bB	373bC	377bB	394aAB	378C
High tunnel	Mid-spring	382bB	390bAB	422aA	408aA	401A
Field/Plastic	Mid-spring	375bB	370bC	377bB	401aAB	381C
Field/no plastic	Mid-spring	383aAB	399aA	386aB	390aB	389B
Field/no plastic	Late-spring	397aA	397aAB	374bB	387abB	389B
Mean		381b	386b	387b	396a	
		----------------Sucrose----------------				
High tunnel	Early-spring	63aA	46bB	33cC	44bB	46B
High tunnel	Mid-spring	37abC	35bC	40abC	45aB	39C
Field/Plastic	Mid-spring	50aB	49aB	39bC	50aAB	47B
Field/no plastic	Mid-spring	59bAB	48cB	73aA	43cB	56A
Field/no plastic	Late-spring	58aAB	59aA	59aB	55aA	58A
Mean		53a	48b	49b	48b	

Field/Plastic	Gardensoy31	9732bc	7323bc
	Gardensoy41	8788bc	6110bcd
	Mooncake	10906bc	6947bc
	Randolph	23883a	17522a
Field/no plastic	Gardensoy31	4884c	2092cd
	Gardensoy41	9163bc	5334bcd
	Mooncake	3822bc	822d
	Randolph	14310ab	10287ab

Table 4. Protein and sucrose content of seeds of edamame cultivars planted at different times and in different production systems. For each seed quality attribute, values in a row followed by same lower case letter are not different (P=0.05), For each seed quality attribute, values in a column followed by same upper case letter are not different (P=0.05).

4. CONCLUSIONS

There is great potential to increase edamame supply in Virginia using season extension techniques. Use of high tunnel and plastic covered seed-beds when ambient conditions are unfavorable for root growth creates localized soil temperatures that allow for improved plant growth. Planted early in the high tunnel, where greenhouse effects increase air temperatures allow early plant growth and early maturity. This crop will be harvested off-season during early summer and producers may access markets when vegetable edamame supply is low and a potential for higher prices exist. A longer period of harvesting time and supply of fresh beans can be achieved with selection of appropriate mix of different MG edamame varieties and production systems. However, economic analysis should be done to determine the viability and profitability of the enterprise that incorporates multiple production techniques. Collaborative work needs to be done with food packaging entities and consumer markets to determine the edamame supply and market demand dynamics so that timely production can be done.

ACKNOWLEDGEMENT

We would like to thank the Sustainable Agriculture Research and Extension (SARE) for funding this project.

REFERENCES

1. Fehr WR, Caviness CE, Burmood DT, Pennington JS (1971) Stage of development descriptions for soybeans, Glycine max (L.) Merrill. Crop Sci 11:929-931.

2. Miles CA, Chen C (2000) Washington State University Edamame trials. WashStUnivCoopExt, Vancouver, WA.

3. Kelley KM, Sanchez ES (2005) Accessing and understanding consumer awareness of and potential demand for Edamame. Hort Sci 40:1347-1353.

4. Johnson D (2000) Edamame: Westerners develop a taste for Japanese soybean. EngTechnolSustainable World 7:11-12.

5. Zhang L, Kyei-Boahen S (2007) Growth and yield of vegetable soybean (Edamame) in Mississippi. HortTechnology 17:26-31.

6. Pao S, Ettinger M, Khalid M, Mebrahtu T, Mullins C (2008) Microbiological quality of frozen "Edamame" (vegetable soybean). JFood Safety 28:300-313.

7. Montri DN, Kelley KM, Sanchez ES (2006) Consumer interest in fresh, in shell Edamame and acceptance of Edamame-based patties. HortSci 41:1616-1622.

8. Park HK, Kim KH, Oh YJ, Kim SC (2001) Selection of vegetable soybean and development of year-round production system in Korea.

9. Bec S, Pfeiffer T, Slone D (2005) Production system for extending the harvest time frame of fresh-market Edamame in Kentucky.

10. AOAC (2005) Official Method 990.03. Protein (Crude) in Animal Feed, Combustion Method, in Official Methods of Analysis of AOAC International.(18thedn),AOAC International, Arlington, VA, pp: 30-31.

11. AOCS (2005) Official Methods and Recommended Practices of the American Oil Chemists Society, Approved procedure Am 5-04 Rapid determination of Oil/Fat utilizing high temperature solvent extraction. Urbana, IL. American Chemical Society.

12. Johansen HN, Glitso V, Knudsen KEB (1996) Influence of Extraction Solvent and Temperature on the Quantitative Determination of Oligosaccharides from Plant Materials by High- Performance Liquid Chromatography. JAgricFood Chem 44: 1470-1474.

13. Waterer D (2003) Yield and economics of high tunnel for production of warm-season vegetable crops. HortTechnology 13:339-343.

14. Carey EE, Jett L, Lamon WJ Jnr, Nennich T, Orzolek MD, et al. (1991)Horticultural Crop Production in High Tunnels in the United States: A Snapshot.HortTechnology 19:37-43.

15. Duppong LM, Hatterman-Valenti H (2005) Yield and quality of vegetable soybean cultivars for production in North Dakota. HortTechnology 15:896-900.

16. Carson LC, Freeman JH, Zhou K, Welbaum G, Reiter M (2011) Cultivar evaluation and lipid and protein contents of Virginia-grown edamame. HortTechnology 21:131-135.

17. Li Y, Du M, Zhang QY, Wang G, Hashemi M, et al. (2013) Correlation and path coefficient analysis for yield and components of vegetable soybean in northeast China. Legume Research 36:284-288.

18. Rao MSS, Bhagsari AS, Mohamed AI (2002) Fresh green seed yield and seed nutritional traits of vegetable soybean genotypes. Crop Sci 42: 1950-1958.

19. Li Y, Du M, Zhang QY, Wang G, Hashemi M, Liu X (2012) Greater differences exist in seed protein, oil, total soluble sugar and sucrose content of vegetable soybean genotypes [Glycine max (L.) Merrill] in Northeast China. AJCS 6:1681-1686.

20. LiY, Du M, Zhang Q, Wang G, Hashemi M, et al. (2014) Planting date influences fresh pod yield and seed chemical compositions of vegetable soybean. HortScience 49:1376-1380.

21. Kumar V, Rani A, Goyal L, Pratap D, Billore SD, et al. (2011) Evaluation of vegetable- type soybean for sucrose, taste-related amino acids, and isoflavones contents. International Journal of food properties 14:1142-1151.

22. Song J, Liu C, Li D, Gu Z (2013) Evaluation of sugar, free amino acid, and organic acid compositions of different varieties of vegetable soybean (Glycine max [L.] Merr). Industrial Crops and Products 50: 743-749.

Chapter 9

Measurement of the light absorption and scattering properties of onion skin and flesh at 633 nm

Weilin Wang, Changying Li*

Bio-Sensing and Instrumentation Laboratory, College of Engineering, The University of Georgia, 712F Boyd Graduate Studies, Athens, GA 30602, USAa

HIGHLIGHTS

- Optical properties of four types of onions were measured at 633 nm.
- Onion skins have significantly higher absorption and scattering coefficients than flesh.
- Onion skins and flesh are high-albedo biological materials.
- The light propagation in multi-layer onion tissues was modeled by Monte Carlo simulations.

ABSTRACT

Understanding the optical properties of onion tissues is essential to applying optical methods for onion quality inspection. This study estimated the optical properties of dry skin, wet skin, and flesh of red, Vidalia sweet, white, and yellow onions at the wavelength of 633 nm. The total diffuse reflectance, total transmittance, and collimated transmittance of single-layer onion tissues were measured by spectroscopic systems. Based on the measured data, the absorption coefficient μa and the reduced scattering coefficient μ'_s of onion tissues were calculated using the inverse adding-doubling method. The results indicated that the dry and wet skins had significantly higher μa and μ'_s than the flesh at 633 nm. For both skins and flesh, the μa varied between cultivars, while the differences of the μ'_s between cultivars were less profound. All types of onion tissues were high-albedo materials at 633 nm. Using the calculated optical properties, Monte Carlo simulations were performed to model the light propagation in 25 different scenarios of multi-layer onion tissues for four cultivars, respectively. The results showed that the incident light at 633 nm would lose 99% of its energy within 6 layers in any of the simulated scenarios, and the light penetrated more layers in the sweet onions than in the other three cultivars. This work provided fundamental understanding of the optical properties of onion tissues and the light propagation in onion bulbs at 633 nm. The investigation of the onion optical properties will be extended to a broader spectrum in the future.

KEYWORDS

Onion; Optical properties; Monte Carlo; Inverse adding-doubling; Scattering; Absorption

1. INTRODUCTION

Onion (*Allium cepa* L.) is an important fresh vegetable which had an approximate production value of $1 billion in the United States in 2012 (USDA, 2013). Mature onions are well ripened bulbs that have concentric layers of dry skins and flesh scales. The flesh scales of onions are edible parts while outer dry skins can be used to make natural food color additive (Block, 2010). Onion skins were also found to be good sources of other by-products such as flavonoids (Ko et al., 2011). The health benefits of onion consumption, such as antifungal, antibacterial, antitumor, antiinflammatory, and antithrombotic, have been widely recognized and advertised in recent decades (Griffiths et al., 2002 and Corzo-martíez et al., 2007).

Nowadays, onion buyers have high expectation of quality. In addition to the nutritional values of onions, other factors such as the appearance, flavor, and existence of defects are also important to consumers (Griffiths et al., 2002). Thus, it is critical to monitor all key onion quality factors throughout the onion harvest and marketing chain. Currently, although some onion packing houses have employed automated systems to sort onions based on the size, other quality properties such as disease infection and surface blemishes have to be evaluated by trained human inspectors, which is often inefficient and subjective. Particularly, human visual inspection is not capable of evaluating the internal quality attributes of onions. Thus, it is necessary to develop more versatile nondestructive techniques to improve the efficacy and efficiency of onion quality inspection.

Among various sensing techniques, optical methods such as imaging, spectroscopy, and spectral imaging are attractive to food industry because they are fast, safe, nondestructive, and low-cost (Zude, 2008 and Cubero et al., 2011). Several optical techniques were reported for onion quality inspection. Birth et al. (1985) applied near-infrared spectroscopy to predict dry matter content of intact onions, and Wang et al. (2012) employed spectral imaging technology to differentiate sour skin-infected onions from healthy ones. The main principle of these techniques was to observe the interaction between the incident light and the onion, which was determined collectively by the optical properties of onion tissues and the properties of the incident light.

Onions are biological tissues and their key optical properties include: the absorption coefficient (μa), scattering anisotropy (g), scattering coefficient (μs), and the refractive index (n) (Tuchin, 2007). The g and μs could be combined into one parameter called reduced scattering coefficient ($\mu'_s=\mu_s(1-g)$). These optical properties are requisite parameters in the radiative transport equation to describe the light reflection, propagation, and attenuation on/in the onion tissue. If the optical properties of onion tissues are known, mathematical methods (such as Monte Carlo simulations) can be applied to numerically model the light-onion interaction, which could provide useful references for designing appropriate optical systems to monitor onion quality. In addition, the optical properties of onion tissues could be indicative to their physical/chemical properties. Therefore, to make optical measurements effectively, it is important to understand and quantify the optical properties of the onion tissues. However, although the physical and chemical attributes of onions have been widely investigated (Maw et al., 1996, Abhayawick et al., 2002 and Rodríguez Galdón et al., 2009), the key optical properties (μa and μs) of onion tissues have never been reported.

Various methods have been reported for measuring the optical properties of biological tissues in the biomedical area (Cheong et al., 1990 and Tuchin, 2007). These methods can be classified into *"ex vivo"* and *"in vivo"* categories in terms of the way of the sample preparation (Kim and Wilson, 2011). The *ex vivo* method, which requires samples of certain shapes and sizes, is often based on the transmittance and reflectance measurements. Then, the optical properties of the measured tissue can be estimated by using computational methods such as the adding-doubling (Prahl et al., 1993 and Pickering et al., 1993) or Monte Carlo (Wang et al., 1995) in their inverse forms. The *in vivo* method is often based on the approximations of radiative transport theory by using the diffuse reflectance or backscattering measurements directly made on the sample in the spatial, temporal, or frequency domains (Kim and Wilson, 2011). In the past decade, the optical properties of a number of food products have been investigated by using *ex vivo* methods such as the apple (Saeys et al., 2008) and mandarin (Fraser et al., 2003), or by *in vivo* methods, such as milk and juice (Qin and Lu, 2007), apple (Qin and Lu, 2008 and Qin and Lu, 2009), and beef (Xia et al., 2007). Generally, the *ex vivo*method is used to obtain the knowledge of the optical properties of a food product, while the *in vivo* method could be used for online inspection of food quality.

Despite the difference, both *ex vivo* and *in vivo* methods require a collimated incident light with a very small beam size (Tuchin, 2007). For this reason, the coherent laser has been considered as one of the standard light sources for measuring the optical properties of tissues. As a relatively inexpensive and reliable light source, laser has short emission time and emits monochrome light beam with high intensity. Due to these advantages, many laser-based optical systems have been reported to measure the optical properties of food products (Cubeddu et al., 2001 and Fraser et al., 2003) or to predict certain quality parameters of food items by measuring their scattering profiles under the laser irradiation (Lu and Peng, 2007, Qing et al., 2008 and Romano et al., 2011).

This study was aimed to quantitatively measure the optical properties of the tissues of four common types of onions (red, sweet, white, and yellow) using a coherent laser. Specific objectives of this research were to:

- Estimate the absorption coefficient μa, the reduced scattering coefficient μ'_s, and the scattering anisotropy g of the onion tissues using laser and integrating shpere based systems.
- Compare the optical properties of the onion skin and flesh.
- Evaluate the differences of the optical properties between four cultivars of onion tissues.
- Model the light propagation in multi-layer onion tissues using Monte Carlo simulation.

2. MATERIALS AND METHODS

2.1. Sample preparation

Onions of four common types: red (cv. Salsa), Vidalia sweet (cv. Century), white (cv. White Cloud), and yellow (cv. Granero) were used in this study. The sweet onions (short-day onions) were harvested from the state of Georgia, USA in May, 2011 and others (long-day onions) were harvested in the state of Idaho, USA in October, 2011. Onions were stored 2–4 months in a cold storage room (2 ± 1 °C) in the Vidalia Onion Research Lab at the University of Georgia. Onions were taken out from the cold room 2–3 h prior to the test and a box fan was used to remove the condensed moisture on the onion surface under room temperature (22 ± 1 °C).

Three types of onion tissues were examined: dry outer skin, wet outer skin, and flesh. The dry outer skin is the onion outer scale(s) dried during harvesting and storage, which protects flesh scales from pathogens in the surrounding environment. It has to be noted that the drying of the outer scales is a slow process accompanied by the biochemical changes of pigments and carbohydrate compositions of cell walls of onion tissues (Brewster, 2008). Thus, in most cases, there is an onion scale whose status is at the intermediate stage between the dry skin and flesh scale, which is called wet outer skin. For the sake of brevity, we used dry skin to represent the onion dry outer skin, and wet skin to indicate the onion wet outer skin in the following sections of this paper.

In each cultivar, 20 large size onion bulbs (76–102 mm in diameter) were tested. All onions were manually selected and inspected to be disease-free and with intact dry skins. Onion bulbs were cut in half longitudinally (from neck to root). In each half of the onion, a set of onion tissues were cut from the equatorial area. At each set of onion tissues, the surface dry skin, the wet skin, the first layer of flesh scale, and the second layer of flesh scale were selected. Two flesh scales instead of one were examined so that the difference between the two layers of scales could be evaluated. Then, selected onion scales were cut into square pieces (30 × 30 mm) and the flesh samples were shaved to slabs using a razor blade. In total, 640 pieces of onion tissue samples (4 cultivars × 20 onions × 2 sample sets × 4 pieces of tissues) were measured.

2.2. Thickness and refractive index of the sample

The thickness (*d*) of each sample was measured using an electronic micrometer (model 35-025, iGaging, San Clemente, CA, USA) with an accuracy of ±0.001 mm in the central area of the tissue. For onion flesh samples, two 1-mm thick glass slides were used to sandwich the sample so that the metal tip of the micrometer could not cut into the tissue, and the thickness of the glass slides were deducted from the measurement later. Three replicates were made for each sample and the average value was reported. The refractive indices (*n*) of samples were measured by a Leica Abbe refractometer (model Abbe Mark II, Reichert, Inc. Depew, NY, USA). For each cultivar, eight pieces of dry skins and eight pieces of flesh tissues were randomly selected and measured. The refractive index of dry skins were measured by directly placing the sample at the prism of the refractometer and using distilled water as the contact liquid. The refractive index of the onion flesh was estimated by testing the onion juice extracted from the tissue. For the dry skin and flesh samples, each of them was measured three times. Then, for each onion group, the mean refractive index was calculated and used in the later calculations.

In practice, it is difficult to measure the refractive index for some wet skin samples by using the Abbe refractometer since they were too thick to be measured as a film but not fleshy enough to be measured by using its juice. It was also observed that, depending on its moisture content, the refractive index of the wet skin of an onion was either close to the one of its dry skin or close to that of its flesh. Thus, based on their thickness values, the refractive indices of onion wet skins were approximated individually to those of dry skins (thin as dry skin) or those of flesh (thick as flesh).

2.3. Reflectance and transmittance measurements

The spectroscopic system (Fig. 1A–C) was assembled to measure the total reflectance, total transmittance, and collimated transmittance spectra of onion samples at 633 nm. The system mainly consisted of an integrating-sphere (model 4P-GPS-060-SF, Labsphere, North Sutton, NH, USA) and a Vis-NIR spectrometer (model USB4000, Ocean Optics, Dunedin, FL, USA). Based on the availability of the device, a low power helium-neon (HeNe) laser (0.8 mW, 633 nm) (model HRP008-1, Thorlabs, Newton, NJ, USA) was used in this study. This type of laser has been widely used for studying the optical properties of biological tissues (Tuchin, 2007 and Kim and Wilson, 2011).

The interior wall of the integrating sphere is coated by Spectraflect material that has a reflectivity of 98% at 633 nm. The internal diameter of the sphere is 152.4 mm and it has four ports (25.4 mm in diameter) at 0°, 90°, 180°, and the north pole. The aperture to surface area ratio of the sphere is less than 3%. The ports at 0° and 180° were used as the entrance and exit, respectively. An optical fiber with a diameter of 400 μm and numerical aperture of 0.37 (model M32L02, Thorlabs, Newton, NJ, USA) was used to deliver the light to the spectrometer from the port at 90° of the sphere. The laser head was placed on a kinematic V-clamp mount (model C1503, Thorlabs, Newton, NJ, USA) so that its position can be precisely adjusted at the horizontal and vertical axes. To align the laser and the sphere, two circular cross targets (25.4 mm in diameter) were printed on transparent films and placed on the entrance and the exit ports, respectively. Then, the positions and orientations of the laser head and the integrating sphere were adjusted until the incident laser beam can go through the center spots of both the entrance and the exit ports. The wavelength accuracy of the spectroscopic system was calibrated by using a pencil style krypton calibration lamp (model 6031, Oriel Instruments, Stratford, CT, USA).

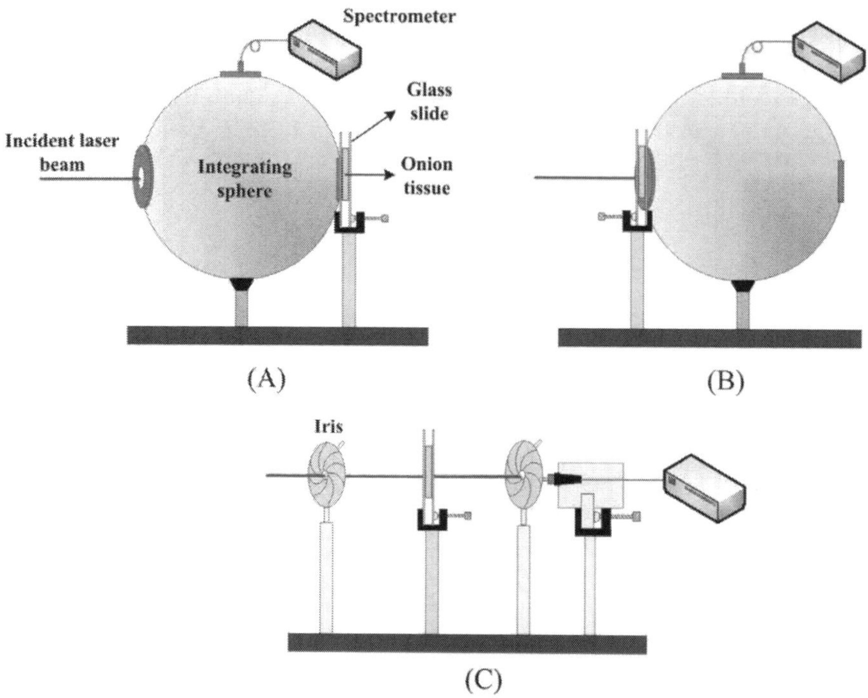

Figure 1. The schematic of the system configurations for measuring the total reflectance (A), the total transmittance (B), and the collimated transmittance (C).

To measure the total reflectance, the sample was placed behind the exit port while the entrance port was open (Fig. 1A). To measure the total transmittance, the sample was placed in front of the entrance port and the opposite exit port was blocked by a Spectraflect plug (Fig. 1B). The reference signal (*Ref*) was measured by using the setup in Fig. 2A, but replacing the onion sample with a certified 99% diffuse reflectance Spectralon target (model AS-00158-060, Labsphere, North Sutton, NH, USA). The dark signal (*D*) of the system was measured by covering all ports of the integrating sphere. Then, the total diffuse reflectance (*Rraw*) and the total transmittance (*Traw*) were converted to the relative total reflectance (*R*) and relative total transmittance (*T*) by:

$$R = \frac{R_{raw} - D}{Ref - D} \quad (1)$$

$$T = \frac{T_{raw} - D}{Ref - D} \tag{2}$$

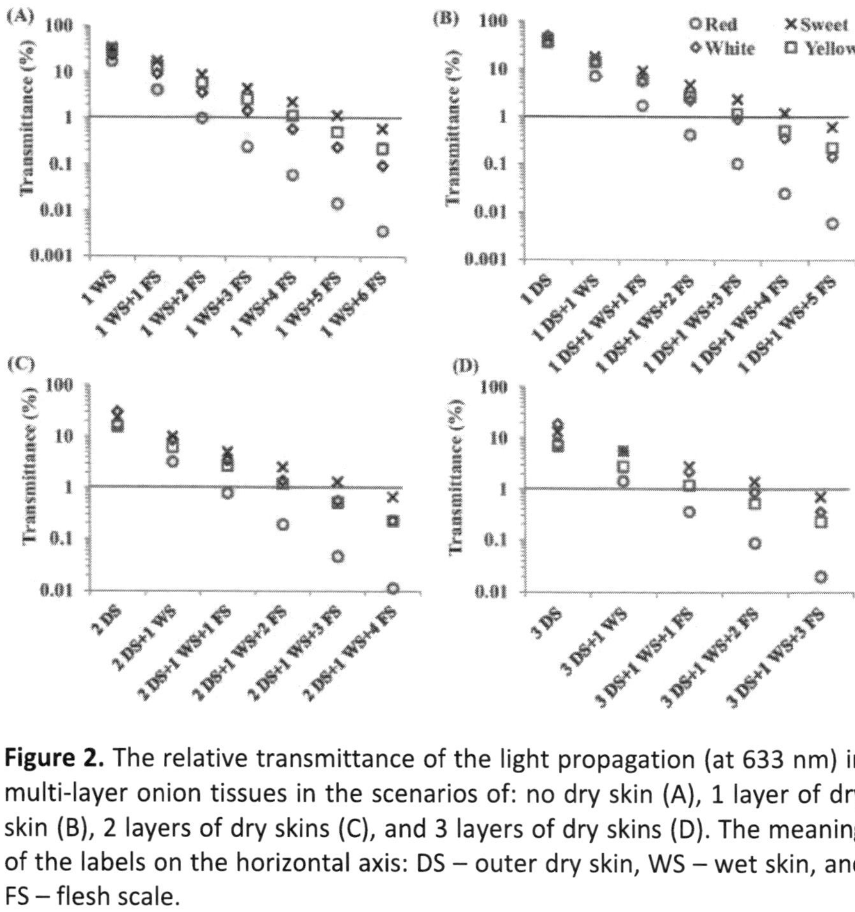

Figure 2. The relative transmittance of the light propagation (at 633 nm) in multi-layer onion tissues in the scenarios of: no dry skin (A), 1 layer of dry skin (B), 2 layers of dry skins (C), and 3 layers of dry skins (D). The meaning of the labels on the horizontal axis: DS – outer dry skin, WS – wet skin, and FS – flesh scale.

The collimated transmittance was measured following the set-up recommended by Prahl (2011) (Fig. 1C). The laser beam was aligned with the center of the detecting optic fiber. An iris was placed between the laser head and the sample to adjust the intensity of the laser beam to avoid saturation of the spectrometer. Another iris in front of the detector was used to block the scattered light from entering the detector. The relative collimated transmittance (*Tc*) was obtained by:

$$T_c = \frac{Tc_{raw} - Dark}{Tc_{ref} - Dark} \quad (3)$$

where *Tcref* was the original light intensity of the incident laser beam and *Tcraw* was the intensity of the laser beam transmitted through the onion tissue sample. The dark signal (*Dark*) of the spectrometer was measured when the probe of the spectrometer was covered by a black cap.

In all measurements, the onion tissue was sandwiched by two pieces of 70 × 70 × 2 mm (width × height × thick) borosilicate glass slides (model BOROFLOAT 33, SCHOTT North America, Inc. Louisville, KY, USA). BOROFLOAT glass was used because of its good homogeneity and high transmittance (>90%) at the visible range. The refractive index of the BOROFLOAT 33 glass is 1.47 at 633 nm. A U-shape clamp and a C-shape clamp were used to apply slight pressure on the glass slides so that the onion tissue formed a flat slab. The whole system was enclosed in a dark chamber covered by a black cloth. All measurements were performed at room temperature (22 ± 1 °C).

2.4. Estimating the absorption and scattering coefficients

The inverse adding-doubling (IAD) program provided by Prahl (2011) was used to calculate the absorption and scattering coefficients for each onion tissue sample. The IAD method has been widely used for estimating the optical properties of biological tissues due to its flexibility and reliability in measuring tissues with various albedos and optical depths (Tuchin, 2007). The fundamental principle of the IAD method is to iteratively solve the stationary (steady-state) radiative transfer equation based on the measured reflection and transmittance values of the sample (Prahl et al., 1993). As the mathematical description of continuous wave light propagation in a medium, the stationary radiative transfer equation (Tuchin, 2007) can be described as:

$$\frac{\partial I(r,\hat{s})}{\partial s} = -(\mu_a + \mu_s)I(r,\hat{s}) + \frac{\mu_s}{4\pi}\int_{4\pi} p(s,\hat{s})I(r,\hat{s})d\Omega \quad 4)$$

where the μa is the absorption coefficient, μs is the scattering coefficient, $I(r,\hat{s})$ is the average flux density at the point *r* along the direction $\hat{s}$, dΩ is the unit solid angle for the direction $\hat{s}$, $p(s,\hat{s})$ is the phase function of the angular distribution of the scattered light, which describes the probability of the polar deflection angle of the photon after each scattering event. The phase function used in all IAD calculations was Henyey-Greenstein (HG) phase function (Prahl et al., 1993), described as $\rho(\theta)=(1/4\pi)\ (1-g^2)/(1+g^2-2g\cos\theta)^{3/2}$. The HG phase function approximates the angular distribution of scattering in biological tissues depending on the mean cosine of scattering angle (the scattering anisotropy coefficient $g = \langle\cos\vartheta\rangle$) (Tuchin, 2007).

Using the IAD program, μ_a and μ'_s can be calculated from the measured *R* and *T* values, with a fixed *g* value. The *g* value can also be estimated by the IAD method when the collimated transmittance T_c is included. In our work, since the anisotropy of onion tissues was unknown before this study, it was first estimated by using the IAD program with three inputs: T, R, and T_c. A small part of data sets (42 out of 640) were removed because of the invalidity of T_c (either T_c was greater than T or the sum of T_c and *R* was equal to or greater than 1). When the g had been estimated, the IAD program with two inputs of Tand R were used to calculate μ_a and μ'_s again. Once g, μ_a, and μ'_s were estimated, $\mu_s$was calculated by $\mu_s=(\mu'_s/(1-g))$.

It should be noted that the IAD program is often prone to error in estimating the anisotropy due to the difficulty of measuring *Tc* accurately (Prahl, 2011). In this study, to compensate the individual bias in the estimated g values, the mean g of each group of onion tissues was calculated and then was used for all samples in the group in later calculations. The more accurate quantifications of anisotropies of onion tissues could be made by using other specific measuring approaches in the future, such as measuring the angular distribution of the scattered light by rotating a collimated detector around the sample enclosed in a cylinder tank with index-matching liquid (Kim and Wilson, 2011).

Analysis of variance (ANOVA) was used to evaluate the differences between four types of onion tissues and between different onion cultivars. The significance of the differences between the means of the measured optical properties of different groups were evaluated by using Tukey's HSD (honest significant difference) tests at the significance level of 0.05. All ANOVA tests were conducted by using the GLM (general linear models) procedure in SAS (v9.2, SAS Institute Inc. Cary, NC, USA).

2.5. Validation procedure

The system was verified by using the phantoms made from a standard scattering material (Intralipid-20%, Sigma Aldrich, St. Louis, Missouri, USA) and two absorption dyes (Nigrosin and Naphthol Green B, Sigma Aldrich, St. Louis, Missouri, USA). In the validation test, the Intralipid-20% was first diluted to one percent by distilled water, and then the absorbers were added to make three different types of liquid solutions. Liquid solutions were added into a quartz glass cuvette with a 10 mm light path to make the phantoms. The cuvette had 40 × 40 mm side walls which can cover the entire sampling port of the integrating sphere. The *T* and *R* of the phantom were measured by the system respectively (5 replicates for each type of phantom). Then, the μ_a and μ'_s of the phantoms were estimated by the IAD program using $g = 0.73$, in which the g was calculated by using the empirical equation ($g(\lambda) = 1.1 - 0.58\lambda$) established by Van Staveren et al. (1991). The estimated values were compared with each other and with the values reported by Ninni et al. (2011).

2.6. Monte Carlo simulation of light propagation in multi-layer onion tissues

Monte Carlo simulations were performed to model the light propagation in multi-layer onion tissues. The "MCML" program provided by Wang et al. (1995) was used for the simulation. Typically, a mature onion bulb has 1–3 layers of dry skins, one layer of wet skin, and a number of flesh scales. Thus, onion models with multiple layers were simulated with the presence of 0–3 layers of dry skins. All examined scenarios had one layer of wet skin and 1–6 flesh scales, which were gradually included in the model. The simulations were stopped at the seventh layer since light rarely can penetrate deeper than that. In each Monte Carlo simulation, 500,000 photons were launched, and the spatial resolution of radial distance and tissue depth was 0.01 mm. The number of grids for the radial and tissue depth was 1000. To simplify the simulation, the onion scales were assumed to be slabs with the mean thickness value listed in Table 1 and the tiny gaps between the inside scales of onions were ignored.

Table 1. The thickness and refractive indices (mean ± SD) of onion tissues in four cultivars.

	Tissue type	Red onion	Sweet onion	White onion	Yellow onion
Thickness (mm)	Dry skin	0.071 ± 0.019	0.087 ± 0.033	0.083 ± 0.025	0.083 ± 0.023
	Wet skin	0.827 ± 0.566	0.468 ± 0.517	0.638 ± 0.802	0.391 ± 0.489
	First flesh	2.833 ± 0.554	3.941 ± 0.734	3.708 ± 0.984	2.962 ± 0.684
	Second flesh	3.317 ± 0.398	3.995 ± 1.109	4.049 ± 0.745	3.392 ± 0.833
Refractive index (n)	Dry skin	1.3345 ± 0.0006	1.3371 ± 0.0074	1.3345 ± 0.0004	1.3337 ± 0.0005
	Flesh	1.3450 ± 0.0007	1.3410 ± 0.0011	1.3413 ± 0.0003	1.3413 ± 0.0015

3. RESULTS

3.1. Thickness and refractive indices of onion scales

The thickness of most onion dry skins tested in this study was 0.07–0.09 mm and the thickness of onion flesh tissue was 3–5 mm (Table 1). As expected, the average thickness values of onion wet skins (0.391–0.827 mm) were in-between those of the dry skin (0.071–0.087 mm) and the flesh (2.833–4.049 mm). On average, the dry skins of the tested red onions were about 15% thinner than those of other cultivars, and the flesh scales of red and yellow onions were about 15–20% thinner than those of sweet and white onions.

Generally, the measured refractive indices of onion flesh (1.341–1.345) were in agreement with the values (1.343–1.351) reported by Foskett and Peterson (1950). For either onion dry skin or onion flesh, the ratio of SD to mean of the refractive index was very small (<0.5%). No statistical difference was observed in flesh or dry skin between cultivars in their refractive indices with the exception of red onion flesh. Since the difference was minute (≈0.003), it was ignored. The averaged values (1.335 for dry skin and 1.342 for flesh) were used for all samples. A threshold value 0.5 mm was arbitrarily selected based on the histogram of the thickness values of the tested wet skins. If a wet skin was thicker than 0.5 mm, the refractive index value for flesh (1.342) was used for it; otherwise, the value of dry skin (1.335) was used.

3.2. System validation

On average, the estimated μ_s' of the phantom showed 3.1% difference to the μ_s' value of Intralipid 20% reported by Ninni et al. (2011). The results of the validation tests showed that the system successfully separated the μ_a and the μ_s' for the three types of phantoms. The estimations of the μ_a and μ_s' for each type of phantom had standard deviations of 1.6–6.7% (on average 3.2%). Based on the two tailed *t* -tests at the significance level of 0.05, the mean μ_s' of three types of phantoms were not significantly different from each other, while their mean μ_a were significantly different ($p < 0.001$). Overall, the accuracy of the system in estimating the μ_a and μ_s' was comparable to other reported studies (Pickering et al., 1993). As discussed by Pickering et al. (1993), the error of the IAD estimation could be caused by the light loss at the rim of the sample and by the interference between the glass slides.

3.3. Scattering anisotropy

The estimated anisotropies of onion tissues at 633 nm (Table 2) varied from 0.35 to 0.73 in different groups, which were smaller than those of regular human tissues (0.8–1) (Tuchin, 2007) and close to those of apples (0.6–0.8) (Saeys et al., 2008). The anisotropies of onion dry skins were lower than those of flesh, except for yellow onions. No significant difference was found between the first and the second flesh scales. The onion wet skins did not show consistent patterns in different cultivars. For red and white onions, the anisotropies of their wet skins were significantly higher than those of dry skins, while not significantly different from those of flesh scales. For sweet and yellow onions, however, the differences between the anisotropies of the wet skins and the dry skins and flesh were not distinctive. This inconsistency can be explained by the data shown in Table 1: the wet skins of red and white onions were thicker than those of sweet and yellow onions, which indicated that the physical characteristics of their wet skins could be more similar to those of flesh scales than those of sweet and yellow onions. These results indicated that the anisotropies of onion tissues could be significantly different between cultivars and between different types of tissues. Therefore, the mean anisotropy of each type of onion tissue in each cultivar was computed and used in later calculations. Since the anisotropies of the first and second flesh scales were not significantly different, they were averaged and used in later computations.

3.4. Absorption and reduced scattering coefficients

Table 3 shows the means and standard deviations of the estimated μ_a and μ'_s values of onion tissues. Onion dry skins had strongest light absorption at 633 nm in red onions (μa = 1.97 × 10^3 m^{-1}), which was significantly higher than those of the other cultivars. The colorless dry skins of white onions showed the lowest μ_a value (0.5 × 10^3 m^{-1}), which was significantly lower than that of yellow onions. The difference between the μ_a of sweet and white onions was not statistically significant. The average μ'_s of dry skins were from 18.53 × 10^3 m^{-1} (white onion) to 22.47 × 10^3 m^{-1} (yellow onion) at 633 nm. The μ'_s of the dry skin of red, sweet, white, and yellow onions were about 10, 25, 37, and 16 times of their corresponding μ_a, respectively. No significant difference was observed between theμ'_s of different cultivars.

Table 2. The means and standard deviations of the estimated anisotropies of onion tissues at 633 nm.

	Tissue type	Red onion	Sweet onion	White onion	Yellow onion
g	Dry skin	$0.589 \pm 0.023^{A,b}$	$0.474 \pm 0.034^{B,b}$	$0.371 \pm 0.035^{B,b}$	$0.545 \pm 0.046^{AB,a}$
	Wet skin	$0.720 \pm 0.025^{A,a}$	$0.589 \pm 0.045^{AB,ab}$	$0.509 \pm 0.032^{BC,a}$	$0.421 \pm 0.047^{C,ab}$
	First flesh	$0.686 \pm 0.015^{A,a}$	$0.727 \pm 0.036^{A,a}$	$0.511 \pm 0.025^{B,a}$	$0.499 \pm 0.045^{B,ab}$
	Second flesh	$0.691 \pm 0.016^{A,a}$	$0.599 \pm 0.042^{AB,ab}$	$0.560 \pm 0.028^{B,a}$	$0.354 \pm 0.037^{C,b}$

Upper case letters in the superscripts indicate the significant differences between the four onion cultivars ($p < 0.05$), and lower case letters indicate the significant differences between the four types of onion tissues ($p < 0.05$).

Table 3. The means and standard deviations of the absorption coefficients μ_a and the reduced scattering coefficients μ'_s of onion tissues at 633 nm.

	Tissue type	Red onion	Sweet onion	White onion	Yellow onion
μ_a ($\times 10^3$ m^{-1})	Dry skin	$1.974 \pm 0.819^{A,a}$	$0.760 \pm 0.460^{C,a}$	$0.501 \pm 0.485^{C,a}$	$1.371 \pm 1.107^{B,a}$
	Wet skin	$0.599 \pm 0.841^{A,b}$	$0.305 \pm 0.316^{AB,b}$	$0.284 \pm 0.378^{B,b}$	$0.325 \pm 0.375^{AB,b}$
	First flesh	$0.117 \pm 0.051^{A,c}$	$0.043 \pm 0.020^{B,c}$	$0.036 \pm 0.018^{B,c}$	$0.033 \pm 0.022^{B,b}$
	Second flesh	$0.093 \pm 0.028^{A,c}$	$0.044 \pm 0.023^{B,c}$	$0.032 \pm 0.010^{B,c}$	$0.033 \pm 0.023^{B,b}$
μ'_s ($\times 10^3$ m^{-1})	Dry skin	$19.733 \pm 8.396^{A,a}$	$19.075 \pm 8.326^{A,a}$	$18.480 \pm 7.524^{A,a}$	$22.476 \pm 15.81^{A,a}$
	Wet skin	$1.988 \pm 2.829^{B,b}$	$3.756 \pm 3.994^{AB,b}$	$4.424 \pm 4.39^{A,b}$	$5.573 \pm 4.044^{A,b}$
	First flesh	$0.616 \pm 0.194^{A,b}$	$0.152 \pm 0.155^{B,c}$	$0.579 \pm 0.319^{A,c}$	$0.659 \pm 0.270^{A,c}$
	Second flesh	$0.600 \pm 0.194^{AB,b}$	$0.253 \pm 0.164^{C,c}$	$0.479 \pm 0.247^{B,c}$	$0.657 \pm 0.300^{A,c}$

Upper case letters in the superscripts indicate the significant differences between the four onion cultivars ($p < 0.05$), and lower case letters indicate the significant differences between the four types of onion tissues ($p < 0.05$).

As illustrated in Table 3, the μ_a and μ_s' of the onion wet skins were in-between those of dry skins and flesh, respectively. The average μ_a of wet skins were about 20–60% of those of the corresponding dry skins, and 5–10 times higher than those of flesh. The average μ_s' of wet skins was about 10–25% of that of dry skins, and 2–18 times higher than that of flesh. Statistical data suggested that the onion wet skin had distinct optical properties from the dry skin and the flesh scale in all cultivars. Thus, it should be considered as a separate type of medium in optical models of onions.

The average estimated μ_s' values of onion flesh were from 0.032×10^3 m^{-1} to 0.117×10^3 m^{-1}, which were significantly smaller (13–41 times) than those of the dry skin. The average μ_a of the flesh scale of red onions was significantly higher than that of the other three cultivars, while no significant difference was found between the other three cultivars. Statistical tests did not reveal any significant difference between the first and the second flesh scales of the μ_a in all examined cultivars. The average μ_s' values of onion flesh were from 0.15×10^3 m^{-1} to 0.66×10^3 m^{-1}. The μ_s' of the flesh of red, white, and yellow onions were quite close, which were significantly higher than those of sweet onions. The mean μ_s' of the flesh of red, white, and yellow onions were about 30 times smaller than those of their dry skins, and the mean μ_s' of sweet onions was about 120 times smaller than that of their dry skins. Similar to the μ_a, no significant difference was presented between the mean μ_s' of the first and the second flesh scales of onions.

4. DISCUSSION

4.1. Optical properties of single layer onion tissues at 633 nm

It is well known that for fruits and vegetables, their scattering coefficients (μ_s or μ_s') are mainly determined by the cellular structures of their tissues, and the absorption coefficients are related to their chemical compounds (Zude, 2008). In this study, the tissues of the four tested onion cultivars did not statistically differ from each other in terms of the μ_s' of their dry skins, but had significant differences between the dry skins of red, yellow, and other onions. In this regard, the results suggested that the dry skins of four onion cultivars had similar cellular structures but had different contents of chemical compounds. As for the onion flesh, it is observed that red onions showed higher μ_a than other cultivars and the μ'_s of sweet onions illustrated significant difference from those of other cultivars at 633 nm. In sum, the optical properties of onion dry skin and flesh could be different between certain cultivars.

Results of this study (Table 4) showed that onion dry skins have greater μ_a and μ'_s than flesh. Onion flesh consists of high percentage of moisture (>90%), glucose, fructose, sucrose, protein, ash, *etc.* (Brewster, 2008). All these major chemical compounds in onion flesh do not have strong absorption at 633 nm. Also, it is known that the absolute content of pigments decreases from outer scales to inner scales in onions (Takahama and Hirota, 2000 and Pérez-Gregorio et al., 2010). During the dehydration process of an onion flesh scale to dry skin, with respect to the decrease of its moisture content, the concentration of the pigment in the scale increases (Takahama and Hirota, 2000) and other physical properties such as the density of the scale also change (Hole et al., 2000). Thus, it is reasonable that the dry skin has higher μ_a and μ'_s than the flesh, since it has greater content of pigments and denser cell structure than the flesh scale.

Compared to the onion dry skin and onion flesh, the onion wet skin showed higher standard errors in their μ_a and μ'_s values. The relative standard error of the μ_a and $\mu'_s$of onion wet skins were 30–100% higher than those of onion dry skins or flesh. This can be explained by the physiological process of onion wet skins in drying. The formation of an onion skin is a slow process affected by many conditions such as the storage time and the relative humidity of the environment. As a result, the physical and chemical properties of onion wet skin could be very different at different stages of drying. Therefore, the optical properties of onion wet skins, intermediate between those of onion dry skins and flesh scales, could have large variations. For this reason, although in this study the μ_a and μ'_swere significantly different between certain tested cultivars, it might not be conclusive that the optical properties of the wet skins of these cultivars are significantly different. The $\mu_a$and μ'_s of onion wet skins reported in this article could only represent the situations of the onions tested in this study.

Overall, the estimated μ_a (0.5×10^3–1.97×10^3 m^{-1}) of onion dry skins were at the same magnitude as those of apple skins (<2×10^3 m^{-1}) reported by Saeys et al. (2008) at 633 nm. The estimated mean μ'_s of onion dry skins (18.48×10^3–22.48×10^3 m^{-1}) were several times higher than those of apple skins (about 3.5×10^3–4×10^3 m^{-1}). The μ_a andμ'_s of onion flesh at 633 nm were at the same order of magnitude with those of other fruits and vegetables (kiwifruit, peach, pear, plum) presented by Qin and Lu (2008).

Table 4. The average absorption coefficient (μa), scattering coefficient (μs), anisotropy (g), thickness (d), refractive index (n), and optical depth (τ) of the single-layer onion tissues of red, Vidalia sweet, white, and yellow onions.

Cultivar	Tissue type	μ_a ($\times 10^3$ m^{-1})	μ_s ($\times 10^3$ m^{-1})	g	d (mm)	n	τ
Red	Dry skin	1.974	47.987	0.589	0.071	1.335	3.547
	Wet skin	0.599	7.104	0.72	0.827	1.335	6.370
	Flesh	0.105	1.9505	0.688	3.075	1.342	6.321
Sweet	Dry skin	0.76	36.243	0.474	0.087	1.335	3.219
	Wet skin	0.305	9.137	0.589	0.468	1.335	4.419
	Flesh	0.044	0.593	0.663	3.968	1.342	2.526
White	Dry skin	0.501	29.385	0.371	0.083	1.335	2.481
	Wet skin	0.284	9.014	0.509	0.638	1.335	5.932
	Flesh	0.034	1.137	0.536	3.879	1.342	4.542
Yellow	Dry skin	1.371	49.355	0.545	0.083	1.335	4.210
	Wet skin	0.325	9.617	0.421	0.391	1.335	3.887
	Flesh	0.033	1.166	0.427	3.177	1.342	3.809

Similar to other biological tissues, onion dry skins and flesh are high-albedo biological media (albedo is defined as $\mu'_s/(\mu_a+\mu'_s)$) at 633 nm. It is known that the biological materials must be scattering dominated to apply the diffusion theory to estimate their optical properties nondestructively (Zude, 2008). Thus, our results supported the feasibility of nondestructively measuring the optical properties of onions at 633 nm by using the diffusion theory, such as the diffuse reflectance or backscattering.

The variances shown in the results of this study were most probably caused by the biological variations (moisture content, pigment content, firmness, *etc.*) of the samples. There were a few other factors that could affect the accuracy of this study. First is the systematic error introduced by the spectroscopic system and the IAD algorithm. In addition, based on our validation tests, the intensity of the laser beam presented a 2.1% deviation in 100 measurements, which could be another error source. Moreover, real onion tissues, particularly onion dry skins, are heterogeneous media with distinguishable textures. However, as a general rule in tissue optics, in this study the onion tissues were assumed being homogeneous to apply the IAD estimation.

Based on the measured optical properties of onion tissues, the light loss caused by absorption and scattering in single layer onion tissues could be estimated. For instance, assuming onion dry skins are pure absorption material (ignoring any scattering effect), based on the definition of μ_a, the probability for photons at 633 nm to survive after traveling 1 mm in the dry skin of red onions and white onions would be 13.81% ($e-\mu_a*d=e-1.98*1$) and 60% ($e-0.5*1$), respectively. Onion bulbs often have 1–3 layers of 0.02–0.1 mm thick dry skins (Brewster, 2008). Thus, for μ_a at this level (0.5×10^3–2×10^3 m^{-1}), the light loss in the dry skins of an onion could approximately be about 1% ($1-e-0.5*0.02$) to 45% ($1-e-1.98*0.3$) depending on its cultivar and the number and thickness of its dry skins. Similarly, if only considering the scattering, using an anisotropy value of 0.5, even in a thin (0.02 mm) dry skin, the approximate light loss caused by the scattering at 633 nm could be higher than 50% ($1-e-(\mu'_s/(1-g))*d= 1-e-(18.53/(1-0.5))*0.02=0.52$). Considering the thickness of the onion, the light absorption and scattering effects at 633 nm in single layer onion tissues can be indicated by their optical depths ($\tau = (\mu a + \mu s) \times d$) (Table 4).

4.2. Light propagation in multi-layer onion tissue at 633 nm

Results of the Monte Carlo simulations revealed that the laser light had different transmission rates in different onion cultivars at 633 nm. Based on the definition of the penetration depth of 1% light transmittance (Fraser et al., 2001), the laser light beam generally penetrated the deepest in Vidalia sweet onions and the least in red onions, while the light transmission in white and yellow onions were in-between (Fig. 2). Overall, the laser beam was not able to penetrate more than 6 onion layers regardless of onion cultivar and layer type. It should be noted that the transmission rate was not only determined by the attenuation factor (combination of absorption and scattering coefficients), but also affected by the thickness of each layer in each cultivar.

The onion dry skin greatly attenuated the propagation of the light in onion tissues in all cultivars. Although onion dry skins were much thinner than flesh scales, the attenuation effect of light from the dry skin was comparable to that from a flesh scale since the dry skin has greater absorption and scattering coefficients than the flesh, which were in accordance with our experimental observations in the total transmittance and collimated transmittance measurements. According to the results of simulations, by adding one, two, and three layers of dry skin(s) to the baseline scenario (one wet skin and one flesh scale) in red onions, the transmission rate of light was reduced from 4.1% to 1.7%, 0.8%, and 0.4%, respectively. After passing three dry skin layers and one wet skin layer of red onions, the light lost 99% of its energy and only a few photons could reach the flesh scales. Thus, in the optical measurements of onions, the onion dry skin should be considered and handled appropriately.

It should be noted that no air gap between onion scales was considered in our simulations. In practice, for some onions, there could be thin air gaps between onion scales. For instance, if the outermost dry skin of the onion bulb is loose, there could be a layer of air between the outermost dry skin and the scale beneath it. The additional air layer could reduce the penetration depth. For example, in the last multi-layer onion model shown in Fig. 2D, when a thin layer of air ($d = 0.036$, $\mu s = 10 \times 10^{-6}\ m^{-1}$, $\mu a = 0.5 \times 10^{-6}\ m^{-1}$) was included between the first and second dry skins, the total diffuse reflectance increased 2.37%, the total absorption and transmission decreased 2.37% and 0.003% (of the overall energy of the initial light), respectively. Therefore, the results of our models could have overestimated the light penetration depth for those onions with air gaps.

5. CONCLUSIONS

Our study revealed that onion dry skin and wet skin had significantly higher absorption and reduced scattering coefficients than onion flesh at 633 nm. Different cultivars of onions could be significantly different in terms of the μ'_s of the flesh and the μ_a of both the flesh and dry skin. Our data also indicated that onion tissues are scattering dominated (high-albedo) biological materials. Monte Carlo simulations of light propagation in multi-layer onion tissues showed that the low power laser at 633 nm would lose 99% of its energy within 6 layers of onion tissues in any of the 25 simulated scenarios because of the substantial light absorption and scattering in onion tissues. The results suggested that the transmittance measurement at 633 nm of an intact onion is not likely and the reflectance or interactance measurements could be more suitable. For the applications that intend to measure the characteristics of onion flesh by using optical methods, it is important to handle onion outer skins properly.

This study is the first effort to measure the light absorption and scattering characteristics of onion tissues, which provides quantitative evidences to better design optical systems for onion quality evaluation. However, this study only investigated the optical properties of onion tissues at one wavelength and future research is needed to expand our understanding to a wider spectral range.

ACKNOWLEDGEMENTS

This work was funded by the USDA NIFA Specialty Crop Research Initiative (Award No.2009-51181-06010). The authors gratefully acknowledge Dr. Ron Gitaitis, Dr. Howard Schwartz, and Dr. Krishna S. Mohan for offering onion samples, Dr. Elizabeth L. Andress for providing the Abbe Refractometer to this study. The assistance from Mr. Anthony M. Dean during data collection was also sincerely appreciated.

REFERENCES

1. Abhayawick, L., Laguerre, J.C., Tauzin, V., Duquenoy, A., 2002. Physical properties of three onion varieties as affected by the moisture content. Journal of Food Engineering 55 (3), 253–262.

2. Block, E., 2010. Garlic and Other Alliums: The Lore and the Science. Royal Society of Chemistry, Burlington House. Piccadilly, London, UK.

3. Birth, G.S., Dull, G.G., Renfroe, W.T., Kays, S.J., 1985. Nondestructive spectrophotometric determination of dry matter in onions. Journal of the American Society for Horticultural Science 110 (2), 297–303.

4. Brewster, J.L., 2008. Onions and Other Vegetable Alliums. Cabi Series. CABI, Cambridge, MA, USA.

5. Cheong, W.F., Prahl, S.A., Welch, A.J., 1990. A review of the optical properties of biological tissues. IEEE Journal of Quantum Electronics 26 (12), 2166–2185.

6. Corzo-martíez, M., Corzo, N., Villamiel, M., 2007. Biological properties of onions and garlic. Trends in Food Science & Technology 18 (12), 609–625.

7. Cubeddu, R., D'Andrea, C., Pifferi, A., Taroni, P., Torricelli, A., Valentini, G., RuizAltisent, M., Valero, C., Ortiz, C., Dover, C., et al., 2001. Time-resolved reflectance spectroscopy applied to the nondestructive monitoring of the internal optical properties in apples. Applied Spectroscopy 55, 1368–1374.

8. Cubero, S., Aleixos, N., Moltó, E., 2011. Advances in machine vision applications for automatic inspection and quality evaluation of fruits and vegetables. Food and Bioprocess Technology 4, 487–504.

9. Foskett, R.L., Peterson, C.E., 1950. Relation of dry matter content to storage quality in some onion varieties and hybrids. Proceedings of the American Society for Horticultural Science 55, 314–318.

10. Fraser, D.G., Jordan, R.B., Künnemeyer, R., McGlone, V.A., 2003. Light distribution inside mandarin fruit during internal quality assessment by NIR spectroscopy. Postharvest Biology and Technology 27 (2), 185–196.

11. Fraser, D.G., Künnemeyer, R., McGlone, V.A., Jordan, R.B., 2001. Near infrared (NIR) light penetration into an apple. Postharvest Biology and Technology 22 (3), 191–194.

12. Griffiths, G., Trueman, L., Crowther, T., Thomas, B., Smith, B., 2002. Onions – a global benefit to health. Phytotherapy Research: PTR 16 (7), 603–615.

13. Hole, C., Drew, R., Gray, D., 2000. Humidity and mechanical properties of onion skins. Postharvest Biology and Technology 19 (3), 229–237.

14. Kim, A., Wilson, B.C., 2011. Measurement of Ex Vivo and In Vivo Tissue Optical Properties: Methods and Theories, 2nd Ed. Springer, Netherlands, pp. 267–319 (Chapter 8).

15. Ko, M.-J., Cheigh, C.-I., Cho, S.-W., Chung, M.-S., 2011. Subcritical water extraction of flavonol quercetin from onion skin. Journal of Food Engineering 102 (4), 327–333.

16. Lu, R., Peng, Y., 2007. Development of a multispectral imaging prototype for real-time detection of apple fruit firmness. Optical Engineering 46 (12), 123201.

17. Maw, B.W., Hung, Y.C., Tollner, E.W., Smittle, D.A., Mullinix, B.G., 1996. Physical and mechanical properties of fresh and stored sweet onions. Transactions of the ASAE 39 (2), 633–637.

18. Ninni, P., Di Martelli, F., Zaccanti, G., 2011. Intralipid: towards a diffusive reference standard for optical tissue phantoms. Physics in Medicine and Biology 56 (2), 21–28.

19. Pérez-Gregorio, R.M., Garca-Falcón, M.S., Simal-Gándara, J., Rodrigues, A.S., Almeida, D.P., 2010. Identification and quantification of flavonoids in traditional cultivars of red and white onions at harvest. Journal of Food Composition and Analysis 23 (6), 592–598.

20. Pickering, J.W., Prahl, S.A., van Wieringen, N., Beek, J.F., Sterenborg, H.J.C.M., van Gemert, M.J.C., 1993. Double-integrating-sphere system for measuring the optical properties of tissue. Applied Optics 32 (4), 399.

21. Prahl, S.A., van Gemert, M.J.C., Welch, A.J., 1993. Determining the optical properties of turbid media by using the adding-doubling method. Applied Optics 32 (4), 559–568.

22. Prahl, S., 2011. Everything I think you should know about Inverse Adding-Doubling. Oregon Medical Laser Center, Manual of the Inverse Adding-Doubling program, http://omlc.ogi.edu/software/iad/ (retrieved 01.10.12).

23. Qing, Z., Ji, B., Zude, M., 2008. Non destructive analyses of apple quality parameters by means of laser-induced backscattering imaging. Postharvest Biology and Technology 48 (2), 215–222.

24. Qin, J., Lu, R., 2007. Measurement of the absorption and scattering properties of turbid liquid foods using hyperspectral imaging. Applied Spectroscopy 61, 388–396. Qin, J., Lu, R., 2008. Measurement of the optical properties of fruits and vegetables using spatially resolved hyperspectral diffuse reflectance imaging technique. Postharvest Biology and Technology 49 (3), 355–365.

25. Qin, J., Lu, R., 2009. Monte Carlo simulation for quantification of light transport features in apples. Computers and Electronics in Agriculture 68, 44–51.

26. Rodríguez Galdón, B., Tascón Rodríguez, C., Rodríguez Rodríguez, E.M., Díaz Romero, C., 2009. Fructans and major compounds in onion cultivars (Allium cepa). Journal of Food Composition and Analysis 22 (1), 25–32.

27. Romano, G., Nagle, M., Argyropoulos, D., Müller, J., 2011. Laser light backscattering to monitor moisture content, soluble solid content and hardness of apple tissue during drying. Journal of Food Engineering 104 (4), 657–662.

28. Saeys, W., Velazco-Roa, M.A., Thennadil, S.N., Ramon, H., Nicolaï, B.M., 2008. Optical properties of apple skin and flesh in the wavelength range from 350 to 2200 nm. Applied Optics 47, 908–919.

29. Takahama, U., Hirota, S., 2000. Deglucosidation of quercetin glucosides to the aglycone and formation of antifungal agents by peroxidase-dependent oxidation of quercetin on browning of onion scales. Plant and Cell Physiology 41 (9), 1021–1029.

30. Tuchin, V.V., 2007. Tissue Optics: Light Scattering Methods and Instruments for Medical Diagnosis, 2nd ed. SPIE Press, Bellingham, WA, USA.

31. USDA, 2013. U.S. Onion Statistics United States Department of Agriculture, http:// usda.mannlib.cornell.edu/ MannUsda/view Document Info. do?documentID =1396 (retrieved 28.03.13).

32. Van Staveren, H.J., Moes, C.J., van Marie, J., Prahl, S.A., van Gemert, M.J.C., 1991. Light scattering in Intralipid-10% in the wavelength range of 400–1100 nm. Applied Optics 30 (31), 4507–4514.

33. Wang, L., Jacques, S.L., Zheng, L., 1995. MCML–Monte Carlo modeling of light transport in multi-layered tissues. Computer Methods and Programs in Biomedicine 47 (2), 131–146.

34. Wang, W., Li, C., Tollner, E.W., Gitaitis, R.D., Rains, G.C., 2012. Shortwave infrared hyperspectral imaging for detecting sour skin (Burkholderia cepacia)-infected onions. Journal of Food Engineering 109 (1), 38–48.

35. Xia, J.J., Berg, E.P., Lee, J.W., Yao, G., 2007. Characterizing beef muscles with optical scattering and absorption coefficients in Vis-NIR region. Meat Science 75 (1), 78–83.

36. Zude, M., 2008. Optical Monitoring of Fresh and Processed Agricultural Crops. CRC Press, Boca Raton, FL, USA.

Chapter 10

Water limitation and rootstock genotype interact to alter grape berry metabolism through transcriptome reprogramming

Mariam Berdeja, Philippe Nicolas, Christian Kappel, Zhan Wu Dai, Ghislaine Hilbert, Anthony Peccoux, Magali Lafontaine, Nathalie Ollat, Eric Gomès & Serge Delrot

[1]Ecophysiologie et Ge´nomique Fonctionnelle de la Vigne, University Bordeaux, ISVV, UMR 1287, Villenave d9Ornon F-33140, France

[2]Department of General and Organic Viticulture, Hochschule Geisenheim University, Geisenheim, Germany

ABSTRACT

Grapevine is a perennial crop often cultivated by grafting a scion cultivar on a suitable rootstock. Rootstocks influence scions, particularly with regard to water uptake and vigor. Therefore, one of the possibilities to adapt viticulture to the extended drought stress periods is to select rootstocks conferring increased tolerance to drought. However, the molecular mechanisms associated with the ability of rootstock/scion combination to influence grape berry metabolism under drought stress are still poorly understood. The transcriptomic changes induced by drought stress in grape berries (cv. Pinot noir) from vines grafted on either 110R (drought-tolerant) or 125AA (drought-sensitive) rootstock were compared. The experiments were conducted in the vineyard for two years and two grape berry developmental stages (50% and 100% veraison). The genome-wide microarray approach showed that water stress strongly impacts gene expression in the berries, through ontology categories that cover cell wall metabolism, primary and secondary metabolism, signaling, stress, and hormones, and that some of these effects strongly depend on the rootstock genotype. Indeed, under drought stress, berries from vines grafted on 110R displayed a different transcriptional response compared to 125AA-concerning genes related to jasmonate (JA), phenylpropanoid metabolism, and pathogenesis-related proteins. The data also suggest a link between JA and secondary metabolism in water-stressed berries. Overall, genes related to secondary metabolism and JA are more induced and/or less repressed by drought stress in the berries grafted on the drought-sensitive rootstock 125AA. These rootstock-dependent gene expression changes are relevant for berry composition and sensory properties.

1. INTRODUCTION

Grapevine is one of the most economically important fruit species worldwide.[1] Grapevine development and ripening are sensitive to environmental factors' and climate change can potentially influence yield, berry composition,[2] and phenology,[3] consequently affecting the viticulture suitability, especially in Mediterranean regions.[4] Among the parameters affected by climate change, water availability drastically limits the distribution and the yield of grapes around the world.[5,6]

The progress made at molecular level concerning gene expression, transcriptional regulation, and signal transduction has improved our understanding of the regulatory networks that control the plant responses to water deficit.[7] Considering grapes, genomic approaches have been used to promote large-scale mRNA expression profiling studies of water and salinity stress.[5,8,9]

Among the different abiotic stresses that affect grapevine, only water deficit, at a moderate level, have been used in a positive way to improve flavor and quality characteristics in grape berries.[10] Significant biochemical changes occurring in berries subjected to water stress may influence berry flavor and composition.[9,11] The total anthocyanin content of water-stressed berries increase[9,11,12] probably due to the upregulation of genes encoding dihydroflavonol-4-reductase (DFR), leucoanthocyanidin dioxygenase (LDOX), UDP-glucose flavonoid 3-O-glucosyltransferase (UFGT), and VvMybA1.13 In parallel with flavonoid induction, sugars also accumulate more under water stress conditions.9,11,12 Furthermore, both abscisic acid (ABA) and sugar signaling might affect berry ripening.11,13, 14, 15, 16 Concerning amino acid biosynthesis, the accumulation of numerous transcripts associated with glutamate and proline (PRO) biosynthesis is upregulated by water deficit.9 Scion cultivars differ in their response to water stress. For instance, under water stress, flavonol content and flavonol synthase transcripts at ripening stage were higher in Chardonnay than in Cabernet Sauvignon.9 The ABA metabolic pathway was affected for both cultivars, but interestingly, under water restriction, ABA concentration was also higher in Cabernet Sauvignon than in Chardonnay.

Currently, almost all vineyards over the world are grafted on rootstocks which are hybrids of three species: *Vitis berlandieri*, *V. riparia*, and *V. rupestris*. In addition to phylloxera resistance, grapevine rootstock genotypes are characterized by the vigor they confer to the scion, which may influence yield,17 grape berry quality,18 berry size, sugar content, anthocyanins, and organic acids19,20 under different irrigation treatments.12 The characterization of the hydraulic features of *Vitis* rootstocks showed a tight correlation between the vigor conferred by the rootstock and the expression of aquaporins (VvPIP2-1, VvPIP2-2) under well-watered conditions.21 Grafting induces massive transcriptional modifications in the shoot apical meristem of the grape scion.22

The selection and breeding of rootstocks improving water use efficiency is one of the key strategies to face climate change. Nevertheless, scion–rootstock interactions, especially under drought stress, are poorly documented at the molecular level. Although some pathways or enzymes affected by water deficit in grape berry have been characterized in the literature, to date no study aimed to evaluate the combined impact of rootstock and water stress on gene expression in the berries in field conditions. In our previous work, two different rootstock genotypes differing in their degree of tolerance to water stress, 110R (drought tolerant) and 125AA (drought sensitive) grafted with the same scion (cv. Pinot noir) were exposed to water limitation in the field, during three years (2009, 2010, and 2011).23 Due to yearly climate variations, the grapes only underwent a significant water stress in 2009 and 2010, and years 2009 and 2010 were therefore selected for detailed metabolite analysis. Under control conditions, predawn leaf water potential (Ψ_{PD}) did not differ between the two rootstocks, and was stable at about −0.10 MPa at harvest for both growing seasons. A weak to moderate water deficit for the two stressed rootstocks was observed during both seasons, with no significant differences of Ψ_{PD} between stressed rootstocks for both years.23 Water stress did not significantly affect berry mass, yield, sugar and organic acids concentration but increased anthocyanin concentration of the berries regardless of the rootstock. Furthermore, the data showed that the berry amino acid concentration at harvest depended on interactions between water supply and rootstocks.

In the present study, a genome-wide transcriptomic approach (microarray) was used to identify ontology categories and genes that were differentially expressed in grape berry grafted on 110R vs. 125AA under water stress, with a special focus on categories linked with grape berry quality.

2. MATERIALS AND METHODS

2.1 Field conditions and sampling

Grape berries from Pinot noir (*Vitis vinifera*) grafted on rootstocks Richter 110 (*V. rupestris* × *V. berlandieri*) and Kober 125AA (*V. riparia* × *V. berlandieri*), were sampled from field-grown vines at two different developmental stages (E-L 35 and E-L 36) according to the modified Eichhorn and Lorenz developmental scale,24 during two growing seasons (2009 and 2010). Vines were trained using a vertical shoot positioning system. The experimental vineyard belongs to the University of Geisenheim (49°59′N, 7°57′W, Germany). Twenty rows of Pinot noir were selected, with an average of 17 plants per row.

Two water conditions were established for the vines grafted on the two rootstocks: control (C) and water stress treatment (S). To impose water stress, half the rows of the vineyard chosen were covered with a removable plastic film from flowering onwards, covering completely the width of the inter-row on both sides of the plants during rainy days. Water stress was applied to the upper part of the vineyard (slope with 6–7% of inclination) and the control was situated at the lower part. Additional details about the training system, ripeness parameters (total soluble solids, titratable acidity, anthocyanins) were published previously.23 Three biological replicates (each including 85 berries from 17 plants per row) for each condition were collected in the vineyard. Sampling was carried out at two berry developmental stages, including 50% veraison when 50% of the berries of a cluster changed their color, and 100% veraison when all berries of a cluster colored. Berries were immediately frozen in liquid nitrogen and stored at −80 °C for further RNA extraction and gene expression analysis.

2.2 RNA extraction

Total RNA from berries of Pinot noir grafted onto 110R and 125AA was extracted as previously described.25 Seeds of each berry were removed before grinding in liquid nitrogen. Total RNA was subjected to DNA digestion with five units of RNase-free DNase I (Promega) for 1 h at 37 °C. RNA integrity was measured at 260 nm with a micro-spectrophotometer (NanoDrop 200C, NanoDrop products, Wilmington, USA) and visualized by electrophoresis on 1.2% agarose gels. RNA quality was also confirmed using a lab-on-chip (2100 Bioanalyzer, Agilent Technologies, Santa Clara, CA, USA). The RNA obtained was then used for quantitative polymerase chain reaction (qPCR) and microarray analysis.

2.3 Quantitative RT-PCR

Quantitative real-time PCR expression analysis was carried out using the CFX96 Real-Time PCR Detection system (Bio-Rad). For each sample, reverse transcription was performed with 2 µg of purified RNA using the Moloney murine leukemia virus reverse transcriptase (Promega) according to the manufacturer's instructions. The cDNA obtained was diluted (1/20) in distilled water. Ten µL reaction mixes were prepared, that included 5 µL of iQTM SYBR Green Supermix (Bio-Rad), 0.2 µM of each primer and 2 µL of diluted berry cDNA. Gene transcripts were quantified upon normalization to *VvGAPDH* and *VvEF1γ* according to.26 Samples from 2009 and 2010 (12 samples for each treatment and rootstock combination) were tested with three technical replicates for each biological sample, and SD values of means were calculated using standard statistical methods. Specific oligonucleotides primer pairs were designed with Beacon Designer 7 software (Premier Biosoft International). Specific annealing of the oligonucleotides was controlled by dissociation kinetics performed at the end of each PCR run. The efficiency of each primer pair was measured on a PCR product serial dilution. Quantitative RT-PCR primer sequences are listed in Table S1.

2.4 Microarray experiment and analyses

The microarrays used were the grape whole-genome microarrays from NimbleGen microarrays 090818 Vitus exp HX12 (Roche, NimbleGen Inc., Madison, WI, USA), representing 29.549 genes based on the 12x genome assembly, using the grapevine V1 gene model predictions (http://genomes.cribi.unipd.it). The chip probe design is available at the following address, http://ddlab.sci.univr.it/FunctionalGenomics. The microarray contains 118.015 probes with an average of four probes per gene. The correspondence between probe identifiers and gene identifiers was obtained from http://genomes.cribi.unipd.it/. All the samples were hybridized following a complete randomized factorial design to take into account the variability through the glass-slide microarrays for broader statistical inference.27 Microarray hybridizations were done for 48 samples (two rootstock × two phenological stages × two years × two treatments × three biological replicates) by the Plateforme Biopuces (Institut National des Sciences Appliquées, Toulouse, France) according to the manufacturer's instructions.

Microarray data were analyzed using the R28 and R/Bioconductor softwares.29 Quality control was done using the ArrayQualityMetrics package,30 and identified outliers were excluded from further analyses. Data were rma normalized31 and differentially expressed genes between stressed and control samples as well as stressed/control differentials between 110R and 125AA rootstocks were identified using the Limma package.32 Differentials with absolute fold changes above 1.5 and *P*-values below 0.05 were considered significant. Significantly affected gene categories were identified using a Wilcoxon rank sum test in R and the MapMan Ontology.33,34 Differentials for the different categories were visualized using PageMan.35 MapMan mappings for the Cribi 12X grapevine genome are based on closest homologs regarding to the *Arabidopsis thaliana* genome. Concerning results presented in the Tables and Supplemental Tables, each sequence corresponding to the ID indicated was manually blasted (blast n) against NCBI database. Microarray data have been deposited in the NCBI GEO database with accession number GSE66391.

3. RESULTS AND DISCUSSION

3.1 Differential gene expression triggered by water stress

The transcriptomic study was performed at two developmental stages that correspond to E-L 35 (50% veraison) and E-L 36 (100% veraison). Indeed, earlier studies reported that veraison corresponds to a key transcriptional reprogramming stage of cellular metabolism that determines most of the phenotypic traits of the ripe berry.36,37 RNA were extracted from berries sampled in 2009 and 2010 on 110R and 125AA plants grown under either control and water stress conditions.

Venn diagrams (Figure 1) shows that at E-L 35, more genes were affected in berries grafted in 110R than in berries grafted on 125AA, while the reverse was true at E-L 36. The proportion of upregulated genes shared for both years ranged between 10 and 30% of the total number of upregulated genes, except for E-L 36 125AA, where this figure drops down to less than 10%. Among the genes differentially expressed for both years, more genes were upregulated than downregulated by water stress, except for E-L 36 125AA. The proportion of downregulated genes shared for both years is 10–20% (Figure 1).

Tables 1 and 2 detail the ontology categories that were identified by Wilcoxon rank sum test as more significantly affected by water stress than other categories at E-L 35 and E-L 36, respectively. For berries grafted on both rootstock genotypes and for both years studied, water stress affected mostly the same categories ('cell wall,' 'lipid metabolism,' 'secondary metabolism,' 'aromatic amino acid metabolism,' 'stress,' 'regulation of transcription,' 'signaling,' etc.). However, this analysis also underlines some specificity due to the rootstock. Indeed, at E-L 35, some categories were specifically affected for 125AA for both years. These included genes related to 'major and minor carbohydrate metabolism,' 'TAG metabolism,' 'aspartate metabolism,' 'flavonoids dihydroflavonols,' 'jasmonate (JA) metabolism,' 'biotic stress,' or 'UDP glucosyltransferases' (Table 1). In contrast, at the same stage, and for both years, transcript abundance of genes involved in 'flavonoids flavonols,' 'polyamine synthesis,' or 'receptor kinases' are more particularly affected in berries grafted on 110R (Table 1). At E-L 36, functional categories specifically affected in berries grafted on 125AA, encompass gene related to 'cellulose synthesis' and 'receptor kinases LRR' (Table 2). In contrast, the categories 'PS light reactions,' 'fatty acid metabolism,' 'flavonols,' 'glutathione S transferases,' and 'ribosomal proteins' were specifically affected in 110R only (Table 2). Altogether, the data show that water stress strongly impacts gene expression in the berries, through major categories that cover cell wall metabolism, secondary metabolism, signaling, stress, and hormones, and that some of these effects strongly dependent on the rootstock genotype.

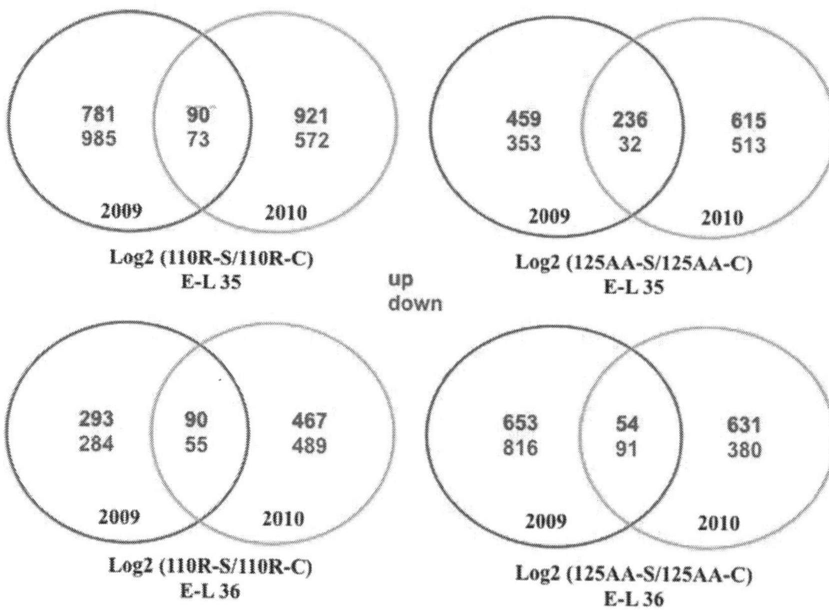

Figure 1. Two-way Venn diagram of the genes differentially expressed at E-L 35 and E-L 36 for the log2(110R-S/110R-C) and log2(125AA-S/125AA-C) conditions in 2009 and 2010.

Some of the genes differentially expressed under water deficit for both years are detailed in supplemental Tables S2 (110R at E-L 35), S3 (125AA at E-L 35), S4 (110R at E-L 36), and S5(125AA at E-L 36).

Table 1. Significantly enriched ontology categories identified by Wilcoxon test at E-L 35 in 2009 and 2010 for Log2 110R-S/110R-C, and Log2 125AA-S/125AA-C respectively

		110R		125AA	
		P-value	*P*-value	*P*-value	*P*-value
Category name	Bin	E-L 35 2009	E-L 35 2010	E-L 35 2009	E-L 35 2010
Major CHO metabolism	2	4.0×10^{-2}	Not significant	1.2×10^{-3}	2.9×10^{-2}
Minor CHO metabolism.raffinose family	3.1	Not significant	Not significant	6.9×10^{-3}	1.1×10^{-2}
Cell wall.cellulose synthesis	10.2	Not significant	Not significant	2.3×10^{-2}	3.0×10^{-2}
Cell wall.degradation	10.6	2.8×10^{-2}	1.3×10^{-2}	1.8×10^{-2}	Not significant
Lipid metabolism.TAG synthesis	11.4	Not significant	Not significant	6.5×10^{-4}	2.3×10^{-3}
Lipid metabolism.lipid degradation	11.9	3.3×10^{-3}	2.0×10^{-2}	4.1×10^{-2}	2.0×10^{-2}
Amino acid metabolism.synthesis.aspartate family	13.1.3	Not significant	Not significant	8.8×10^{-3}	9.7×10^{-3}
Amino acid metabolism.synthesis.aromatic aa	13.1.6	3.7×10^{-2}	1.1×10^{-2}	4.8×10^{-2}	2.3×10^{-5}
Secondary metabolism.isoprenoids	16.1	Not significant	1.7×10^{-2}	4.6×10^{-2}	2.3×10^{-3}
Secondary metabolism.phenylpropanoids. lignin biosynthesis	16.2.1	3.3×10^{-4}	4.0×10^{-2}	4.6×10^{-8}	3.2×10^{-7}
Secondary metabolism.flavonoids. chalcones.naringenin-chalcone synthase	16.8.2.1	2.9×10^{-13}	1.8×10^{-5}	0.0	0.0
Secondary metabolism.flavonoids.di-hydroflavonols.flavonoid 3-monooxygenase	16.8.3.3	Not significant	Not significant	4.0×10^{-2}	1.3×10^{-2}
Secondary metabolism.flavonoids.flavonols	16.8.4	2.4×10^{-2}	7.5×10^{-6}	Not significant	Not significant
Hormone metabolism.jasmonate.synthesis-degradation	17.7.1	Not significant	Not significant	6.9×10^{-3}	1.9×10^{-2}
Stress.biotic	20.1	Not significant	4.9×10^{-4}	4.2×10^{-4}	2.6×10^{-2}
Stress.abiotic	20.2	6.5×10^{-9}	3.0×10^{-4}	1.6×10^{-3}	1.2×10^{-4}
Polyamine metabolism.synthesis	22.1	3.1×10^{-2}	5.2×10^{-3}	Not significant	Not significant
Misc. UDP glucosyl and glucoronyl transferases	26.2	Not significant	Not significant	2.7×10^{-2}	3.6×10^{-4}
RNA.regulation of transcription	27.3	2.4×10^{-2}	4.2×10^{-2}	2.1×10^{-3}	3.2×10^{-3}
Protein.synthesis.ribosomal protein	29.2.1	Not significant	0.0	9.7×10^{-5}	2.2×10^{-4}
Signaling.receptor kinases.wheat LRK10 like	30.2.20	$1.2 \times 10^{-}2$	3.2×10^{-4}	Not significant	Not significant
Development.late embryogenesis abundant	33.2	Not significant	1.7×10^{-2}	$3.9\ 10^{-2}$	1.8×10^{-2}
Not assigned.no ontology. pentatricopeptide (PPR) repeat-containing protein	35.1.5	1.3×10^{-2}	7.3×10^{-20}	7.2×10^{-6}	1.7×10^{-7}

Abbreviations: CHO, carbohydrate; 125AA, Kober 125AA (*Vitis berlandieri* × *Vitis riparia*); 110R, Richter 110 (*Vitis berlandieri* × *Vitis rupestris*).

Table 2. Significantly enriched ontology categories identified by Wilcoxon test at E-L 36 in 2009 and 2010 for Log2 110R-S/110R-C, and Log2 125AA-S/125AA-C respectively

Category name	Bin	110R *P*-value E-L 36 2009	110R *P*-value E-L 36 2010	125AA *P*-value E-L 36 2009	125AA *P*-value E-L 36 2010
PS.lightreaction	1.1	1.3×10^{-2}	0.0	Not significant	Not significant
Cell wall.cellulose synthesis	10.4	Not significant	Not significant	2.8×10^{-2}	1.1×10^{-2}
Cell wall.degradation	10.6	3.7×10^{-2}	7.6×10^{-6}	1.6×10^{-2}	3.1×10^{-2}
Lipid metabolism.FA synthesis and FA elongation	11.1	2.8×10^{-2}	1.9×10^{-3}	Not significant	Not significant
Lipid metabolism.lipid degradation	11.9	2.6×10^{-2}	1.5×10^{-3}	7.5×10^{-3}	7.8×10^{-3}
Secondary metabolism.isoprenoids	16.1	3.0×10^{-2}	3.4×10^{-2}	3.6×10^{-2}	Not significant
Secondary metabolism.phenylpropanoids.lignin biosynthesis	16.2.1	3.2×10^{-4}	3.0×10^{-4}	Not significant	1.4×10^{-9}
Secondary metabolism.sulfur-containing. glucosinolates.degradation	16.5.1.3	2.8×10^{-4}	7.2×10^{-6}	1.4×10^{-4}	8.0×10^{-3}
Secondary metabolism.flavonoids.chalcones. naringenin-chalcone synthase	16.8.2.1	5.6×10^{-4}	1.3×10^{-18}	5.4×10^{-14}	1.9×10^{-10}
Secondary metabolism.flavonoids.anthocyanins	16.8.1	4.4×10^{-4}	9.9×10^{-3}	Not significant	3.8×10^{-2}
Secondary metabolism.flavonoids. dihydroflavonols.flavonoid 3-monooxygenase	16.8.3.3	2.3×10^{-3}	1.0×10^{-2}	1.9×10^{-5}	2.0×10^{-2}
Secondary metabolism.flavonoids.flavonols	16.8.4	6.2×10^{-5}	3.6×10^{-4}	Not significant	Not significant
Secondary metabolism.simple phenols	16.10	1.2×10^{-7}	3.8×10^{-5}	Not significant	1.4×10^{-3}
Hormone metabolism.jasmonate.synthesis-degradation	17.7.1	1.1×10^{-2}	4.9×10^{-2}	1.3×10^{-2}	Not significant
Stress.abiotic	20.2	1.0×10^{-3}	5.5×10^{-6}	6.9×10^{-5}	2.9×10^{-2}
Stress.biotic	20.1	6.1×10^{-3}	2.1×10^{-4}	1.5×10^{-3}	1.2×10^{-7}
Redox	21	8.4×10^{-3}	5.5×10^{-3}	2.2×10^{-2}	Not significant
Misc.glutathione S transferases	26.9	8.2×10^{-3}	4.9×10^{-2}	Not significant	Not significant
RNA.regulation of transcription	27.3	2.7×10^{-2}	4.8×10^{-2}	2.4×10^{-3}	1.6×10^{-2}
Protein.synthesis.ribosomal protein	29.2.1	2.9×10^{-13}	1.9×10^{-2}	Not significant	Not significant
Signaling.receptor kinases.leucine rich repeat XI	30.2.11	Not significant	Not significant	3.6×10^{-2}	2.9×10^{-2}
Signaling.receptor kinases.thaumatin like	30.2.15	1.7×10^{-3}	1.0×10^{-11}	3.2×10^{-8}	1.0×10^{-5}
Signaling.receptor kinases.wheat LRK10 like	30.2.20	1.7×10^{-3}	3.4×10^{-7}	1.3×10^{-15}	3.1×10^{-8}
Development.late embryogenesis abundant	33.2	1.4×10^{-4}	4.6×10^{-2}	Not significant	3.0×10^{-4}
Transport	35.1.5	Not significant	3.9×10^{-3}	2.5×10^{-3}	9.6×10^{-3}

Abbreviations: 125AA, Kober 125AA (*Vitis berlandieri* × *Vitis riparia*); 110R, Richter 110 (*Vitis berlandieri* × *Vitis rupestris*).

3.2 Differential gene expression triggered by water stress between the berries grafted on the two rootstocks

In order to highlight the effect of rootstock under drought stress conditions on the transcriptional response of the berries, the ratios 110R-stressed (S)/110R-control (C) vs. 125AA-S/125AA-C were compared by using the relation log2(110R-S/110R-C) – log2(125AA-S/125AA-C). A positive ratio (log2(110R-S/110R-C) – log2(125AA-S/125AA-C)) means that genes are more induced or less repressed by drought stress in berries grafted on 110R (abbreviated: '110R' condition) in comparison to berry grafted on 125AA (abbreviated: '125AA' condition). Conversely, a negative ratio (log2(110R-S/110R-C) – log2(125AA-S/125AA-C)) means that genes are more induced or less repressed by drought stress in 125AA berries than in 110R berries.

Using Wilcoxon rank sum test, we identified functional categories differentially affected among the genes differentially expressed between 110R and 125AA under drought stress (Table 3). Interestingly, genes involved in secondary metabolism (phenylpropanoid and flavonoids), hormone metabolism (JA), and biotic stress response (pathogenesis-related proteins (PR proteins)) were among the highest differentially expressed genes (P-value < 0.05) for both years (2009 and 2010) and both developmental stages (E-L 35 and E-L 36; Table 3).

The next section will describe and discuss the functional categories cited above. Given that some differences in amino acid composition were described earlier with these samples,23 we will also detail the genes related to amino acid metabolism. Some of the genes differentially expressed in the microarray data were confirmed by qPCR in the same samples (from 2009 to 2010) and results are presented in Table S6. The direction of gene expression changes could generally be repeated by qPCR; however, the magnitude of change was frequently increased in the qPCR data.

3.3 Effects of rootstock genotype on phenolic compounds metabolism under water deficit

The phenolic compounds derived from the shikimic acids pathway are precursors of simple phenols, lignin, flavonoids, isoflavonoids, coumarin, and stilbenes. Polyphenolic compounds play an important role in the quality of grapes and wines.[38] Flavonoids are important determinants of quality in red grapes and wines. Color and taste of red wines are strongly related to the amount of anthocyanins, flavonols, and proanthocyanidins. Many of these phenolics may be beneficial for human health.[39]

Our transcriptomic data agree with previous observations according to which water deficit affects phenylpropanoid metabolism and increases the berry concentration in anthocyanins and other phenolics.[10,11,40] Furthermore, we show that the effects of water deficit on the expression of genes of this family in the berries depend on the rootstock. Indeed, the Pageman representation shows that, overall, genes from the functional category 'secondary metabolism' (more particularly phenylpropanoids and flavonoids) are more induced or less repressed by drought stress in berries grown on the drought-sensitive rootstock 125AA compared to the drought-tolerant rootstock 110R, except for E-L 36 developmental stage in 2009 (Figure 2A).

Both water deficit in the field and postharvest dehydration increase the accumulation of phenylalanine ammonia lyase (*PAL*) transcripts[8,9,41] which may indicate a general activation of the phenylpropanoid pathway. No evidence for any significant differential regulation of *PAL* genes between 125AA and 110R was found. However, other genes involved in the first steps of the phenylpropanoid pathway are differentially expressed between 110R and 125AA under drought stress: a 4-coumarate-CoA ligase (*4CL, Vv18s0001g00290*) and a trans-cinnamate 4-monooxygenase (*C4H, Vv11s0078g00290*) in E-L 35 2010 (Table 5) and E-L 36 2009 (Table 6), respectively. The *C4H* gene is the most differentially affected by water deficit between the two rootstocks due to both its upregulation (1.6X) in 110R stressed berries coupled with a downregulation (2X) in 125AA stressed berries (Table 6; Figure 2A).

In addition to the upstream steps, downstream steps of the phenylpropanoid pathway were also affected. Water deficit accelerates ripening and alters flavonoid biosynthesis, especially the anthocyanin pathway.[11] General biosynthetic genes such as *DFR* (*Vv15s0048g00980*) and *F3H*(*Vv07s0031g01380* and *Vv16s0098g00860*) were differentially expressed between 110R and 125AA in 2009 (Tables 4 and 6): *DFR* was downregulated by drought stress in 110R in comparison to 125AA for E-L 35 (Table 4). At E-L 36, *F3H* genes were more downregulated by drought stress in 125AA than in 110R (Table 6). Besides these general biosynthetic genes, genes related to anthocyanin modification (including *flavonoid 3'5'-*

hydroxylase and *O-methyltransferase*) can also be upregulated in water stressed berries.13 Our results further show that the rootstocks also participated in the control of genes related to anthocyanins modification, more particularly anthocyanin 3-O-glucoside-6'-O-malonyltransferase (*Vv12s0134g00590*), UDP rhamnose-anthocyanidin-3-glucoside-rhamnosyl transferase (*Vv19s0014g02020*, *Vv15s0046g01960*, *Vv00s0218 g00190*) and cyanidin-3-O-glucoside-2-O-glucuronosyl transf erase (*Vv04s0044g015 30*; Tables 4, 5, 6).

Flavonols, which are also products of the flavonoid biosynthetic pathway, contribute to the bitter taste and color of red wine by stabilizing anthocyanin pigments.38 Deficit irrigation moderately affects flavonol synthesis in red grapevines42 and Chardonnay.8,9 Like anthocyanins, changes induced by water deficit affect more the range of flavonols than their global concentration.43 In our study, two genes encoding flavonol modifying enzymes were differentially expressed between 110R and 125AA under drought stress: 8-hydroxyquercetin-8-O-methyltransferase (*Vv12 s0028g02890*) and quercetin-3-O-methyltransferase (*Vv15s0048g 02490*) at E-L 35 2009 (Table 4).

Isoflavonoids are defense compounds mostly found in legumes44 that may have beneficial effects in humans and animals.45 Little information is available related to the involvement of isoflavonoids in grape ripening. IFR6 (isoflavone reductase) which catalyzes the reduction of isoflavones to isoflavonones decreases from veraison to the end of ripening.46 Accumulation of isoflavones is promoted by well-watered conditions and long-term progressive drought significantly decreases isoflavone content,47 but the molecular basis of this response remains elusive. Proteomic studies showed that IFR6 protein is more abundant in the pulp of berries under water-deficit stress,48which is in agreement with its transcript profile.8 Our study also suggests a role for *IFR* in drought stress response since *IFR3* (*Vv07s0031g03070*) in 2009 (Table 4), *IFR2* (*Vv03s0088g00060*), and*IFR6* (*Vv03s0038g04620*) in 2010 (Table 5) are strongly differentially expressed between 110R and 125AA under water deficit at E-L 35.

Table 3. Significantly enriched ontology categories identified by Wilcoxon test for the log2(110R-S/110R-C) – log2(125AA-S/125AA-C) condition that are common for the two developmental stages in 2009 and 2010

Category name	Bin	*P*-value	*P*-value	*P*-value	*P*-value
		E-L 35 2009	E-L 36 2009	E-L 35 2010	E-L 36 2010
Secondary metabolism	16	2.9×10^{-3}	4.8×10^{-10}	9.6×10^{-4}	5.7×10^{-8}
Phenylpropanoids	16.2	4.7×10^{-3}	5.4×10^{-4}	7.9×10^{-3}	4.6×10^{-7}
Flavonoids	16.8	2.0×10^{-2}	6.9×10^{-5}	3.3×10^{-3}	1.0×10^{-5}
Hormones metabolism. Jasmonate					
Synthesis degradation	17.7.1	3.5×10^{-2}	1.5×10^{-2}	2.1×10^{-3}	1.7×10^{-2}
Biotic stress	20.1	5.5×10^{-3}	2.3×10^{-6}	2.26×10^{-6}	4.3×10^{-6}
PR proteins	20.1.7	5.8×10^{-4}	7.4×10^{-3}	9.3×10^{-14}	4.1×10^{-12}

Abbreviation: PR proteins, pathogenesis-related proteins.

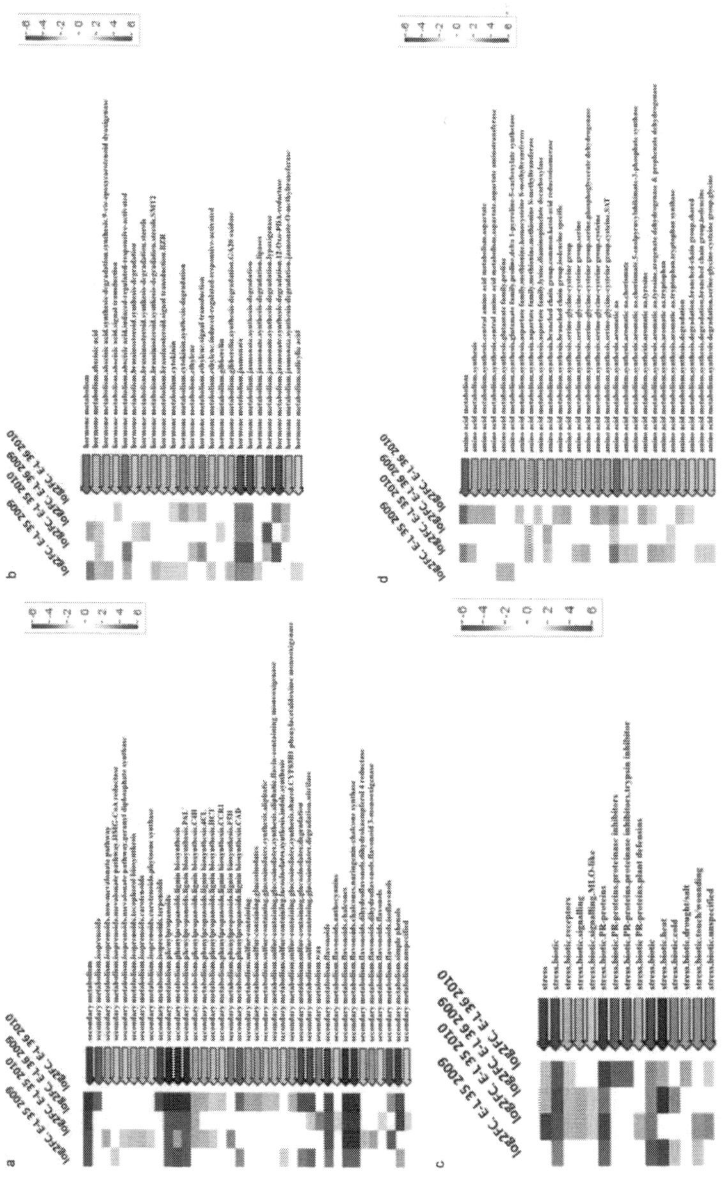

Figure 2. Pageman visualization of MapMan functional categories: secondary metabolism (**A**), hormone metabolism (**B**), stress (**C**), and amino acid metabolism (**D**), enriched in the genes differentially expressed between berries grafted on 110R and berries grafted on 125AA, under drought stress. The degree of enrichment of functional categories in up- and downregulated genes for the log2(110R-S/110R-C) – log2(125AA-S/125AA-C) condition is given by shades of red and blue, respectively.

Proanthocyanidins or condensed tannins are flavan-3-ol oligomers responsible for wine bitterness and astringency. Changes occurring in proanthocyanidins during grape development are complex, involving increases in the degree of polymerization, in the proportion of (-)-epigallocatechin extension units, and in polymer-associated anthocyanins.49 In grapevine, both anthocyanidin reductase (ANR) and leucoanthocyanidin reductase contribute to proanthocyanidins synthesis in leaves, flowers, and in the skin and seeds of the developing fruit.50 In seeds, water deficit does not affect *ANR* transcript levels.8 The slight increases in skin tannin that accompany water deficit51would result more from differential growth sensitivity of the inner mesocarp and the exocarp than from direct effects on phenolic biosynthesis.10 Our study differs somewhat from these earlier data since we found a differential expression of *ANR* (*Vv00s0361g00030*, for E-L 35 2009 and 2010 (Tables 4 and 5)), (*Vv00s0361g00040* for E-L 36 2010 (Table 7)), and *LDOX* (*Vv08s 0105g00380*, for E-L 36 2009 (Table 6)) between 110R and 125AA under water deficit.

Lignins are cell wall phenolic heteropolymers mainly synthesized by the oxidative coupling of p-hydroxycinnamyl alcohol monomers. Lignification and biosynthesis of lignin precursors is upregulated by drought.52 The general phenylpropanoid pathway is branched to the biosynthesis of monolignol by two enzymatic steps catalyzed by cinnamoyl-CoA reductase (CCR) and cinnamyl alcohol dehydrogenase (CAD). CAD enzyme may also be involved in the synthesis of cinnamyl alcohol derivatives responsible for fruit flavor and aroma.53 In the present study, *CAD* genes are less repressed (*Vv13s0019g05260*) or more induced (*Vv18s0001g14910*) by drought stress in 125AA than in 110R berries at E-L 35 in 2009 and 2010, respectively (Tables 4 and 5). Another alcohol dehydrogenase (*SAD, Vv00s0346g00100*) was also more repressed by drought stress in 110R than in 125AA berries at E-L 35 in 2009 (Table 4). A similar expression pattern was observed for other transcripts involved the biosynthesis of monolignols, such as caffeoyl-CoA O-methyltransferase (*CCoAOMT, Vv01s0010g03460*) and caffeic acid 3-O-methyltransferase (*COMT, Vv18s0001g02610*; Table 4). Both *CCoAOMT* and *CAD* transcrits were upregulated during water stress in rice54 and *Citrullus lanatus* sp.55 Another key lignin biosynthesis-related gene, *HCT*(*Vv11s0037g00590, Vv11s0037g00590*), encoding hydroxycinnamoyl-CoA shikimate/quinate hydroxycinnamoyl transferase, is more positively regulated by drought stress in 125AA than in 110R berries at E-L 35 for both years (Tables 4 and 5). HCT leads to the biosynthesis of two major monolignols, namely, the guaiacyl and syringyl units.56 In line with our results, drought treatment caused an early and biphasic expression of *HcHCT* in *Hibiscus cannabinus*.57

Table 4. Selected genes affected in condition E-L 35 (2009) associated with JA biosynthesis, phenylpropanoid metabolism, PR proteins, amino acid metabolism, and transport and polyamines ontology categories

Name	ID	Fold change Log2(110R-S/110R-C) – log2(125AA-S/125AA-C)	*P*-value	Fold change Log2 (110R-S/110R-C)	Fold change Log2 (125AA-S/125AA-C)
Jasmonate synthesis-related genes					
Phospholipase A1	*Vv15s0021g01510*	−1.13	0.04	−0.78	0.35
Linoleate 13S-lipoxygenase	*Vv06s0004g01470*	−1.13	0.01	−0.40	0.56
Phenylpropanoids					
Simple phenols					
Laccase 15	*Vv18s0075g00950*	0.95	0.00	0.40	−0.56
Multicopper oxidase	*Vv18s0164g00110*	−0.65	0.02	−0.46	0.18
Laccase 1a	*Vv08s0007g01910*	−0.75	0.03	−0.69	0.06
Laccase 15	*Vv13s0019g02160*	−0.80	0.01	−0.16	0.64
Laccase 110b	*Vv08s0007g00600*	−0.82	0.02	−0.21	0.61
Laccase 11	*Vv04s0069g00950*	−0.96	0.02	−0.52	0.44
Laccase 1	*Vv17s0000g02590*	−1.06	0.00	−0.45	0.61
Laccase	*W13s0019g0194*	−1.11	0.02	−0.77	0.34
Lignins					
HCT	*Vv11s0037g00590*	−0.75	0.03	−0.47	0.29
Sinapyl alcohol dehydrogenase	*Vv00s0346g00100*	−0.94	0.05	−0.58	0.36
CAD	*Vv13s0019g05260*	−1.02	0.02	−0.96	0.06
CCoAOMT	*Vv01s0010g03460*	−0.67	0.04	−0.41	0.27
COMT	*Vv18s0001g02610*	−0.99	0.02	−0.34	0.65
Flavonoids					
8-hydroxyquercetin 8-O-methyltransferase	*Vv12s0028g02890*	0.68	0.01	0.33	−0.35
Quercetin 3-O-methyltransferase	*Vv15s0048g02490*	1.01	0.01	0.53	−0.48
Anthocyanin 3-O-glucoside-6"-O-malonyltransferase	*Vv12s0134g00600*	0.79	0.00	0.33	−0.46
UDP rhamnose-anthocyanidin-3-glucoside rhamnosyltransferase	*Vv19s0014g02020*	−0.89	0.03	−1.07	−0.18
ANR	*Vv00s0361g00030*	1.07	0.04	0.48	−0.59
DFR	*Vv15s0048g00980*	−1.08	0.02	−0.67	0.41
Isoflavone reductase 3	*Vv07s0031g03070*	−1.76	0.01	−0.89	0.86

PR proteins					
Endochitinase PR4-like	*Vv05s0094g00220*	−1.05	0.05	−0.35	0.70
Amino acid metabolism					
AAAs					
Arogenate dehydratase	*Vv10s0116g01670*	−1.30	0.00	−0.99	0.31
Arogenate dehydrogenase	*Vv09s0002g08030*	0.74	0.04	0.60	−0.15
PAT	*Vv14s0083g00900*	−1.13	0.02	−0.72	0.41
Methionine					
MET6	*Vv10s0042g00970*	−1.02	0.03	−0.72	0.30
Lysine					
Aminotransferase ALD1	*Vv18s0001g04630*	0.75	0.03	0.57	−0.19
Valine. leucine. isoleucine					
Acetolactate synthase	*Vv16s0022g01100*	−1.23	0.02	−0.86	0.37
Amino acid transport					
AAP	*Vv00s0188g00130*	−1.04	0.01	−0.59	0.44
AAP	*Vv04s0008g00230*	−1.41	0.00	−0.71	0.70
Polyamines					
Ornithine decarboxylase	*Vv18s0001g00740*	0.81	0.01	1.08	0.27
S-adenosylmethionine decarboxylase	*Vv14s0083g00580*	0.60	0.01	0.62	0.02

Abbreviations: AAAs, aromatic amino acids; AAP, amino acid permease; ANR, anthocyanidin reductase; CAD, cinnamyl alcohol dehydrogenase; CCoAOMT, caffeoyl-CoA O-methyltransferase; COMT, caffeic acid 3-O-methyltransferase; DFR, dihydroflavonol-4-reductase; HCT, hydroxycinnamoyl-CoA shikimate/quinate hydroxycinnamoyl transferase; JA, jasmonate; 125AA, Kober 125AA (*Vitis berlandieri* × *Vitis riparia*); PAT, glutamate/aspartate-prephenate aminotransferase; PR proteins, pathogenesis-related proteins; 110R, Richter 110 (*Vitis berlandieri* × *Vitis rupestris*).

Table 5. Selected genes affected in condition E-L 35 (2010) associated with JA biosynthesis, phenylpropanoid metabolism, PR proteins, and amino acid metabolism and transport ontology categories

Name	ID	Fold change Log2(110R-S/110R-C) – log2 (125AA-S/125AA-C)	*P*-value	Fold change Log2(110R-S/110R-C)	Fold change Log2(125AA-S/125AA-C)
Jasmonate synthesis-related genes					
Jasmonate-O-methyltransferase	*Vv18s0001g12900*	−1.18	0.01	−0.40	0.78
Phenylpropanoids					
4CL	*Vv18s0001g00290*	−0.63	0.00	−0.28	0.35
Simple phenols	*Vv08s0007g01910*	1.77	0.01	1.25	−0.52
Laccase 1a	*Vv09s0018g00950*	0.94	0.04	1.11	0.17
Laccase	*Vv11s0037g00440*	−1.09	0.04	−0.36	0.66
Lignins	*Vv18s0001g14910*	−1.02	0.00	−0.20	0.89
HCT	*Vv15s0046g01960*	−0.78	0.04	−0.31	0.47
CAD	*Vv00s0218g00190*	−0.65	0.00	−0.06	0.59
Flavonoids	*Vv04s0044g01530*	0.90	0.04	0.24	−0.66
Anthocyanin 3-glucoside rhamnosyltransferase					
Anthocyanin 3-glucoside rhamnosyltransferase	*Vv00s0361g00030*	0.94	0.01	0.69	−0.25
Cyanidin-3-O-glucoside 2-O-glucuronosyltransferase					
ANR	*Vv03s0088g00060*	2.30	0.03	2.34	0.03
Isoflavone reductase 2					
Isoflavone reductase 2	*Vv03s0088g00250*	1.35	0.01	1.37	0.01
Isoflavone reductase 2	*Vv03s0088g00140*	0.90	0.04	1.00	0.10
Isoflavone reductase 6	*Vv03s0038g04620*	0.60	0.03	0.52	−0.08
PR proteins					
Endochitinase PR4-like	*Vv05s0094g00270*	−0.90	0.02	−0.35	0.55
Basic form of pathogenesis-related protein 1	*Vv03s0088g00940*	−0.95	0.00	−0.15	0.79
Amino acid metabolism					
AAAs					
Arogenate dehydrogenase	*Vv09s0002g08060*	0.80	0.01	0.65	−0.15
Chorismate mutase	*Vv01s0010g00480*	0.99	0.03	1.42	0.44
Methionine					
MMT	*Vv18s0089g01290*	0.66	0.03	0.20	−0.46
Amino acid transport					
Amino acid perméase	*Vv04s0008g00230*	1.13	0.03	1.70	0.57
LHT	*Vv00s0302g00030*	0.94	0.02	0.52	−0.42
LHT	*Vv01s0010g01490*	0.74	0.02	0.59	−0.15

Abbreviations: AAAs, aromatic amino acids; ANR, anthocyanidin reductase; CAD, cinnamyl alcohol dehydrogenase; 4CL, 4-coumarate-CoA ligase; HCT, hydroxycinnamoyl-CoA shikimate/quinate hydroxycinnamoyl transferase; JA, jasmonate; 125AA, Kober 125AA (*Vitis berlandieri* × *Vitis riparia*); LHT, lysine/histidine transporter; MMT, methionine S-methyltransferase; PAT, glutamate/aspartate-prephenate aminotransferase; PR proteins, pathogenesis-related proteins; 110R, Richter 110 (*Vitis berlandieri* × *Vitis rupestris*).

Table 6. Selected genes affected in condition E-L 36 (2009) associated with JA biosynthesis, phenylpropanoid metabolism, PR proteins, amino acid metabolism, and polyamines ontology categories

Name	ID	Fold change Log2(110R-S/110R-C) – log2(125AA-S/125AA-C)	*P*-value	Fold change Log2(110R-S/110R-C)	Fold change Log2(125AA-S/125AA-C)
Jasmonate synthesis-related genes					
Jasmonate-O-methyltransferase	*Vv04s0023g02290*	1.06	0.01	0.68	−0.38
Phenylpropanoids					
C4H	*Vv11s0078g00290*	1.55	0.03	0.67	−0.88
Simple phenols					
Laccase 15	*Vv18s0075g01090*	−1.20	0.02	−0.12	1.08
Flavonoids					
Flavonoid 3-hydroxylase	*Vv07s0031g01380*	0.71	0.02	−0.33	−1.04
LDOX	*Vv08s0105g00380*	0.63	0.02	0.05	−0.58
Anthocyanin 3-O-glucoside-6″-O-malonyltransferase	*Vv12s0134g00590*	−0.85	0.02	−0.14	0.71
Flavanone 3-dioxygenase	*Vv16s0098g00860*	0.84	0.02	−0.19	−1.03
Stilbenes					
STS	*Vv16s0100g00770*	1.22	0.00	−0.25	−1.46
STS	*Vv16s0100g01140*	1.07	0.03	0.05	−1.02
PR proteins					
Endochitinase PR4-like	*Vv05s0094g00200*	0.87	0.03	0.22	−0.65
Amino acid metabolism					
Aminotransferase ALD1	*Vv12s0055g00920*	1.24	0.02	−0.46	−1.70
Polyamines					
Ornithine decarboxylase	*Vv18s0001g00740*	1.35	0.01	−0.24	−1.59

Abbreviations: JA, jasmonate; 125AA, Kober 125AA (*Vitis berlandieri* × *Vitis riparia*); LDOX, leucoanthocyanidin dioxygenase; PR proteins, pathogenesis-related proteins; 110R, Richter 110 (*Vitis berlandieri* × *Vitis rupestris*); STS, stilbene synthase; C4H, trans-cinnamate 4-monooxygenase.

Table 7. Selected genes affected in condition E-L 36 (2010) associated with JA biosynthesis, phenylpropanoid metabolism, PR proteins, and amino acid transport ontology categories

Name	ID	Fold change Log2(110R-S/110R-C) – log2(125AA-S/125AA-C)	*P*-value	Fold change Log2(110R-S/110R-C)	Fold change Log2(125AA-S/125AA-C)
Jasmonate synthesis-related genes					
Jasmonate-O-methyltransferase	*Vv04s0023g02240*	−0.75	0.01	−0.22	0.53
Phenylpropanoids					
Simple phenols					
Laccase 15	*Vv18s0075g01090*	−1.19	0.04	−1.03	0.16
Laccase	*Vv18s0117g00480*	−1.76	0.02	−1.37	0.39
Laccase	*Vv00s0444g00010*	−1.95	0.03	−1.51	0.44
Laccase	*Vv18s0117g00450*	−1.99	0.03	−1.49	0.50
Laccase	*Vv18s0075g00610*	−2.01	0.04	−1.52	0.49
Laccase	*Vv18s0075g00780*	−2.43	0.04	−1.84	0.59
Laccase	*Vv00s1212g00020*	−2.43	0.02	−2.08	0.36
Laccase	*Vv18s0075g00670*	−2.48	0.03	−1.87	0.61
Laccase	*Vv00s0731g00010*	−2.83	0.02	−2.41	0.42
Flavonoids					
ANR	*Vv00s0361g00040*	0.51	0.04	0.53	0.02
Stilbenes					
STS	*Vv16s0100g01110*	−0.61	0.01	−0.45	0.16
STS	*Vv16s0100g01190*	−2.16	0.03	−1.60	0.56
PR proteins					
PR-1 protein	*Vv03s0088g00780*	−1.05	0.04	−0.51	0.54
Basic form of pathogenesis-related protein 1	*Vv03s0088g00710*	−1.95	0.02	−0.98	0.97
Amino acid transport					
LHT	*Vv01s0010g02500*	−0.63	0.04	−0.68	−0.05

Abbreviations: ANR, anthocyanidin reductase; JA, jasmonate; 125AA, Kober 125AA (*Vitis berlandieri* × *Vitis riparia*); LHT, lysine/histidine transporter; PR proteins, pathogenesis-related proteins; 110R, Richter 110 (*Vitis berlandieri* × *Vitis rupestris*); STS, stilbene synthase.

Stilbenes are other nonflavonoid compounds present in small quantities in wine.58 Their synthesis increases upon pathogen infection and in response to abiotic stress. Stilbene synthase (STS) is the key enzyme leading to the biosynthesis of resveratrol, which may be beneficial to human health. Water deficit increases the amounts of *STS* transcripts, which suggests an increase in resveratrol accumulation.8 However, little effect of drought was observed on resveratrol concentrations in skins.59 Water deficit increased the accumulation of trans-piceid (the glycosylated form of resveratrol) in Cabernet Sauvignon but not in Chardonnay berries.60 In the present study, four genes encoding STSs were differentially expressed under water stress depending on the rootstock genotype. Two genes (*Vv16s0100g00770, Vv16s0100g01140*) were downregulated in 125AA berries in 2009 at E-L 36 (Table 6) and two others (*Vv16s0100g01110, Vv16s0100g01190*) were strongly downregulated in 110R berries at E-L 36 in 2010 (Table 7).

Finally, several laccase-encoding genes were strongly differentially expressed between 110R and 125AA at both developmental stages for both years (Tables 4, 5, 6, 7). Laccases are widely present in eukaryotes and in prokaryotes.61 In addition to their role in the oxidative polymerization of monolignols, they are involved in the oxidative polymerization of flavonoids.62 Although valued in tea and wine processing, browning reactions are particularly detrimental for fresh fruits.63 Laccases play a role in defending the plant against various biotic and abiotic stresses64 more particularly during acclimation to salinity stress.65 The present study identified several laccases such as *LAC 15, 1a, 1, 110B,* and *11* that were differentially expressed in berries grown on 110R and 125AA stressed rootstocks (Tables 4, 5, 6, 7). Overall, they were found to be more induced or less repressed in 125AA than in 110R berries for both years (Tables 4, 6, and 7).

3.4 Effects of rootstock genotype on hormone metabolism under water deficit

The rootstock genotype may affect JAs metabolism under water stress condition. For both years and both developmental stages, transcripts related to JAs metabolism were significantly more induced or less repressed in 125AA than in 110R berries (except for E-L 36 2009; Figure 2B). JAs is well known to regulate responses to wounding and defense against pathogens but has also been reported to play a role in drought response.66 For instance, JAs were found to cause stomatal closure52 and JA-insensitive mutants (*jar1* and *coi1*) 67,68 show impaired stomatal response to exogenous JAs. There is a cross-talk between JAs and ABA because they use a similar cascade of events to promote stomatal closure.68 Furthermore, a role of JAs in plant response to water deficit has been suggested because this stress induces the expression of several genes that also respond to JAs.69 Microarray analysis in wild-type and *coi1* Arabidopsis plants that were wounded or exposed to water stress reveals a surprisingly large overlap of *COI1*-dependent genes. Moreover, like ABA, JAs levels were increased by drought stress in soybean70 and *Pinus pinaster*.71 The same was observed with rice, for which drought and high salinity increased leaf and root JAs, resulting in the induction of stress-related PR and JAs biosynthetic genes.72 However, these abiotic stress-induced increases in JAs levels were observed only in vegetative tissues. Whether drought conditions increase JAs levels in reproductive organs remains to be determined. Our data emphasize that JAs may play a role in grape berry acclimation during drought stress. Indeed, key JAs biosynthesis genes such as phospholipase A1 (*Vv15s0021g01510*), linoleate 13S-lipoxygenase (*Vv06s0004g01470*), and jasmonate-O-methyltransferase (*Vv18s0001 g129 00, Vv04s0023g02290,Vv04s0023g02240*) were differentially regulated under water deficit between the drought-tolerant rootstock 110R and the drought-sensitive rootstock 125AA (Tables 4, 5, 6, 7).

JA compounds regulate the biosynthesis of many different types of secondary metabolites in different plant species including alkaloids, terpenoids, glucosinolates, and phenylpropanoids.73 For instance, exposure of *Hypericum perforatum* L. suspension culture to JA upregulates the expression of *PAL* and chalcone isomerase (*CHI*) genes. This observation was correlated with an increased production of phenylpropanoids such as phenolic, flavanol and flavonol in JA-treated cells.74 In Arabidopsis, JAs induce anthocyanin accumulation through the upregulation of the 'late' anthocyanin biosynthetic genes *DFR*, *LDOX*, and *UFGT*.75 In grapes, JAs promote the synthesis and accumulation of stilbenes in berry cell cultures76 and during berry development.59 The present work pinpoints a possible link between JAs and phenylpropanoid pathway in grape berry under water-deficit condition.59 Indeed, genes related to JA metabolism and those related to phenylpropanoid/flavonoid metabolism follow a similar expression pattern for both developmental stages and for both years (Figure 2A and 2B). At E-L 35 (2009 and 2010) and E-L 36 (2010), the genes related to these two functional categories are more induced or less repressed in berries grafted on 125AA than on those grafted on 110R (Figure 2A and 2B). At E-L 36 in 2009, those genes were more induced or less repressed in 110R than in 125AA berries (Figure 2A and 2B).

3.5 Effects of rootstock genotype on pathogen-related proteins under water stress

JA is a well-known inducer of PR protein during plant defense responses.77 In grapevine, the most frequent and best characterized defense reactions to fungal infection are the accumulation of phytoalexins and pathogen-related PR proteins.78 The latter accumulate in berries and leaves upon pathogen attack and may contribute to grapevine resistance.79 Besides, due to their resistance to proteolysis and stability at acidic pH, PR proteins are the most abundant class of protein present in wines and adversely affect their clarity and stability.80,81 PR proteins include 17 categories of structurally and functionally unrelated proteins,82 some on which have been detected in grapevine under water stress conditions.83 PR proteins designated as PR 1, 2, 3, 4, 5, 6, 10, and 14 are differentially expressed in several berry tissues.8 The expression pattern of genes related to stress in the present study is shown in Figure 2C. Among them, the transcript abundance of some specific*PR* genes differed between 110R and 125AA under drought stress, e.g. *PR-4* and *PR1* (Tables 4,5, 6, 7). Interestingly, PR4 has been reported to be one of the markers for the JA pathway84 and the subfamily PR-1 may be involved in the response to environmental stresses82 such as drought stress.85

3.6 Effects of rootstock genotype on amino acid metabolism under water stress

The discussion above highlighted the functional categories differentially expressed between 110R and 125AA under drought stress for both developmental stages and both years. Our previous study, describing the effects of rootstock genotype under water stress on berry composition, revealed that specific amino acids accumulated more in 125AA than in 110R berries.23 Even though the 'amino acid metabolism' functional category as a whole is not statistically affected (Table 3), some genes belonging to this category display differential expression pattern between 110R and 125AA under water deficit (Tables 4, 5, 6). Figure 2D shows that transcripts related to 'amino acids synthesis' were more induced or less repressed in 125AA than in 110R berries in drought stress conditions. This observation fits well with our biochemical analysis showing that berries from 125AA accumulated more amino acids than 110R berries.23 Furthermore, genes related to aromatic amino acids (AAAs) (Figure 2D) present the same expression pattern as the ontology categories 'secondary metabolism' (Figure 2A) and 'hormone metabolism jasmonate' (Figure 2B).

First of all, our data suggest that the shikimate pathway is differentially affected between 110R and 125AA under drought stress (Figure 2D). It is well known that exposure of plants to various stresses generally induces the expression of genes encoding enzymes of the shikimate pathway and AAAs pathway. AAAs include phenylalanine (PHE), tyrosine (TYR), and tryptophan (TRP). This pathway also provides precursors for a variety of plant hormones, such as auxin and salicylate, as well as for a very wide range of secondary metabolites (e.g. alkaloids, flavonoids, and lignin)86 with multiple biological functions and environmental responses.87 TRP is a precursor of alkaloids, phytoalexins, and indole glucosinolates as well as the plant hormone auxin, whereas TYR is a precursor of isoquinoline alkaloids, pigment betalains, and quinones.88 PHE is a common precursor of numerous phenolic compounds, including flavonoids, condensed tannins, lignans, lignin, and phenylpropanoid/benzenoid volatiles.89 TYR and PHE are synthesized from arogenate by the key enzymes arogenate dehydrogenase and arogenate dehydratase, respectively, with the relative flux to each being tightly controlled. The partitioning between TYR and PHE is controlled by the sensitivity of arogenate dehydrogenase to feedback inhibition by TYR.90 The present study shows a differential expression between 110R and 125AA at EL-35 concerning arogenate dehydrogenase (*Vv09s0002g08030* and *Vv13s0067g02120* for 2009, *Vv09s0002g08060* for 2010) and arogenate dehydratase (*Vv10s0116g01670* for 2009; Tables 4 and 5). Interestingly, arogenate dehydrogenase and arogenate dehydratase transcripts were respectively more induced and strongly more repressed in 110R berries than in 125AA berries under drought stress (Tables 4 and 5). This observation may partially explain that the phenylpropanoid pathway, which starts from PHE, is more active in 125AA than in 110R berries at E-L 35 under stress (Figure 2A). Arogenate dehydrogenase genes are also differentially regulated in drought-challenged root tissues of two parental genotypes of chickpea.91

Others genes involved in amino acids biosynthesis via the arogenate route are also differentially expressed between 110R and 125AA under drought stress at E-L 35, such as glutamate/aspartate-prephenate aminotransferase (*PAT, Vv14s0083g00900* for 2009) and chorismate mutase (*Vv01s0010 g00480* for 2010; Tables 4 and 5). The same is true for genes related to methionine (MET) metabolism (*MET6, Vv10s0042g00970* and *MMT, Vv18s0089g01290*; Tables 4 and 5). In the literature, there are few and ambiguous data on MET accumulation under drought92 and osmotic stress conditions.93 Our data also reveal a differential expression for amino acid transferase (*ALD1:Vv 18s0001g04630* and *Vv12s0055g00920*; Tables 4 and 6). In *Arabidopsis*, ALD1 generates an amino acid derivative important for the activation of defense signaling, and recombinant ALD1 exhibits aminotransferase activity with a strong substrate preference for lysine.94

No gene directly related to PRO metabolism, well known to play a role in adaptation to drought stress, was differentially expressed between 110R and 125AA. In contrast, our data reveal a differential expression of genes related to polyamines biosynthesis (ornithine decarboxylase, *Vv18s0001 g00740* and S-adenosylmethionine decarboxylase, *Vv14s008 3g00580*; Tables 4 and 6). Polyamines (PAs) are small aliphatic amines implicated in a wide range of developmental processes and environmental stresses.95 Remarkably, drought stress induces changes in PAs amounts that broadly correlate with drought resistance traits.96 PAs have also been reported to play a role in grape ripening.97 Interestingly, genes related to PAs biosynthesis (and more particularly ornithine decarboxylase) may also regulate the entire subset of pathways for glutamic acid (GLU) to arginine (ARG) and to PRO, a common osmolyte.98 Therefore, we can hypothesize that the differences in ARG, PRO, and GLU content between 125AA and 110R under drought stress we observed in our previous study23 could be partially due to a differential transcriptional regulation of ornithine decarboxylase (*Vv18s0001g00740*; Tables 4 and 6).

Finally, under drought stress, the rootstock genotypes also impact genes related to amino acid transport (Tables 4, 5, and 7). Plant amino acid transporters (ATFs) are classified into two superfamilies; the amino acid, polyamine, and choline transport superfamily and the ATF family superfamily.99,100 The ATF superfamily contains five sub-classes of transporters including the amino acid permeases (AAPs), lysine/histidine transporters (LHTs), and the PRO transporters. At E-L 35 in 2009, two *AAPs* (*Vv00s0188g 00130* and *Vv04s0008g 00230*) were upregulated by drought stress in 125AA and downregulated in 110R (Table 4). In contrast, two *LHTs* (*Vv00s0302g 00030* and*Vv01s0010g 01490*) were upregulated in 110R and downregulated in 125AA by drought stress at E-L 35 in 2010 (Table 5).

4. CONCLUSIONS

Although systems biology is increasingly used to study the effects of abiotic stress in plants,101,102no report has yet investigated the effects of rootstock genotype on the transcriptional response of grape berries to water deficit. The present data provide an overall view of the dynamic changes of the grape berry (cv. Pinot noir) transcriptome during drought stress for berries grafted on a drought-tolerant and a drought-sensitive rootstock. Although the variable environmental conditions occurring in 2009 and 2010 resulted in different levels of water stress, the comparative transcriptome analysis highlighted significant rootstock-dependent differences in the response of genes related to secondary metabolism, JA metabolism, and PR proteins for both years and both developmental stages. Among these, the changes occurring in secondary metabolism are of particular interest for selecting the rootstocks and as a consequence influencing grape berry and wine quality. Overall, genes related to secondary metabolism are more induced and/or less repressed by drought stress in the berries grafted on the drought-sensitive rootstock 125AA. Furthermore, our data also suggest that JA may play a role during a weak to moderate water stress, and may be responsible for the subsequent modulation of the secondary metabolism (e.g. stilbenes) of grape berries.

There are many ways in which the genotype of the rootstock may ultimately affect gene expression in the berries. These include water and nutritional signals depending on root extension and activities, vigor, hormone signaling, physical signaling, and macromolecular signals. This comprehensive transcriptomic analysis will be very useful not only to investigate the molecular network controlling secondary metabolism during drought stress but also to explore further the molecular links between JA and secondary metabolism.

ACKNOWLEDGEMENTS

We would like to dedicate this paper in deepest recognition to the memory of Dr. Anthony Peccoux, assistant professor of viticulture at the College of Agriculture, Food and Natural Resources, University of Missouri, who passed away in November 2012. Anthony enthusiastically designed the field trial and collected data during season 2009 in the course of his PhD work. We thank Pr. Hans Schultz and Pr. Manfred Stoll (Hochschule Geisenheim University) for helpful discussions. The PhD work of M. Berdeja was supported by an Erasmus Mundus grant. Part of this work was supported by the Conseil Régional d'Aquitaine (France) and the land of Hesse (Germany).

REFERENCES

1. Vivier MA, Pretorius IS. Genetically tailored grapevines for the wine industry. Trends Biotechnol 2002; 20: 472–478.

2. Jones GV, Davis RE. Using a synoptic climatological approach to understand climate-viticulture relationships. Int J Climatol 2000; 20: 813–837.

3. Webb LB, Whetton PH, Barlow EW. Modelled impact of future climate change on the phenology of winegrapes in Australia. Austral J Grape Wine Res 2007; 1: 165–175.

4. Hannah L, Roehrdanz PR, Ikegami M, Shepard AV, Shaw MR et al. Climate change, wine, and conservation. Proc Natl Acad Sci USA 2013; 110: 6907–6912.

5. Cramer GR, Ergül A, Grimplet J, Tillett RL, Tattersall EA et al. Water and salinity stress in grapevines: early and late changes in transcript and metabolite profiles. Funct Integr Genomics 2007; 7: 111–134.

6. Chaves MM, Zarrouk O, Francisco R, Costa JM, Santos T et al. Grapevine under deficit irrigation: hints from physiological and molecular data. Ann Bot 2010; 105: 661–676.

7. Seki M, Umezawa T, Urano K, Shinozaki K. Regulatory metabolic networks in drought stress responses. Curr Opin Plant Biol 2007; 10: 296–302.

8. Grimplet J, Deluc LG, Tillett RL, Wheatley MD, Schlauch KA et al. Tissue-specific mRNA expression profiling in grape berry tissues. BMC Genomics 2007; 8: 187.

9. Deluc LG, Quilici DR, Decendit A, Grimplet J, Wheatley MD et al. Water deficit alters differentially metabolic pathways affecting important flavor and quality traits in grape berries of Cabernet Sauvignon and Chardonnay. BMC Genomics 2009; 10: 212.

10. Roby G, Harbertson JF, Adams DA, Matthews MA. Berry size and vine water deficits as factors in winegrape composition: anthocyanins and tannins. Austral J Grape Wine Res 2004; 10: 100–107.

11. Castellarin SD, Matthews MA, Di Gaspero G, Gambetta GA.Water deficits accelerate ripening and induce changes in gene expression regulating flavonoid biosynthesis in grape berries.Planta 2007; 227: 101–112.

12. Koundouras S, Hatzidimitriou E, Karamolegkou M, Dimopoulou E, Kallithraka S et al. Irrigation and rootstock effects on the phenolic concentration and aroma potential of*Vitis vinifera* L. cv. Cabernet Sauvignon grapes. J Agric Food Chem 2009; 57: 7805–7813.

13. Castellarin SD, Pfeiffer A, Sivilotti P, Degan M, Peterlunger E et al. Transcriptional regulation of anthocyanin biosynthesis in ripening fruits of grapevine under seasonal water deficit.Plant Cell Environ 2007; 30: 1381–1399.

14. Gambetta GA, Matthews MA, Shaghasi TH, McElrone AJ, Castellarin SD. Sugar and abscisic acid signaling orthologs are activated at the onset of ripening in grape. Planta 2010;232: 219–234.

15. Kuhn N, Guan L, Dai ZW, Wu BH, Lauvergeat V et al. Berry ripening: recently heard through the grapevine. J Exp Bot 2014; 65: 4543–4559, 2014.

16. Lecourieux F, Kappel C, Lecourieux D, Serrano A, Torres E et al. An update on sugar transport and signalling in grapevine. J Exp Bot 2014; 65: 821–832.

17. Stevens RM, Pech JM, Gibberd MR, Walker RR, Nicholas PR. Reduced irrigation and rootstock effects on vegetative growth, yield and its components, and leaf physiological responses of Shiraz. Austral J Grape Wine Res 2010; 16: 413–425.

18. Tandonnet JP, Cookson SJ, Vivin P, Ollat N. Scion genotype controls biomass allocation and root development in grafted grapevine. Austral J Grape Wine Res 2010; 16: 290–300.

19. Ollat N, Tandonnet JP, Bordenave L, Decrocq S, Geny L et al.Vigour conferred by rootstock: hypotheses and direction for research. Bull OIV 2003; 76: 581–595.

20. Cortell JM, Halbleib M, Gallagher AV, Righetti TL, Kennedy JA. Influence of vine vigor on grape (*Vitis vinifera* L. cv. Pinot Noir) anthocyanins. 1.

Anthocyanin concentration and composition in fruit. J Agric Food Chem 2007; 55: 6575–6584.

21. Gambetta GA, Manuck CM, Drucker ST, Shaghasi T, Fort K et al. The relationship between root hydraulics and scion vigour across Vitis rootstocks: what role do root aquaporins play? J Exp Bot 2012; 63: 6445–6455.

22. Cookson SJ, Ollat N. Grafting with rootstocks induces extensive transcriptional re-programming in the shoot apical meristem of grapevine. BMC Plant Biol 2013; 13: 147.

23. Berdeja M, Hilbert G, Dai ZW, Lafontaine M, Stoll M et al.Effects of water stress and rootstock genotype on Pinot Noir berry composition. Austral J Grape Wine Res 2014; 20: 409–421.

24. Coombe BG. Adoption of a system for identifying grapevine growth stages. Austral J Grape Wine Res 1995; 1: 104–110.

25. Reid KE, Olsson N, Schlosser J, Peng F, Lund ST. An optimized grapevine RNA isolation procedure and statistical determination of reference genes for real-time RT-PCR during berry development. BMC Plant Biol 2006; 6: 27.

26. Vandesompele J, De Preter K, Pattyn F, Poppe B, Van Roy N et al. Accurate normalization of real-time quantitative RT-PCR data by geometric averaging of multiple internal control genes. Genome Biol 2002; 18: 7: RESEARCH0034.

27. Yang YH, Speed T. Design issues for cDNA microarray experiments. Nat Rev Genet 2002; 3: 579–588.

28. R Core Team. R: A Language and Environment for Statistical Computing. Vienna: R Foundation for Statistical Computing; 2013. ISBN 3-900051-07-0. http://www.R-project.org/.

29. Gentleman R, Carey V, Bates D et al. Bioconductor: open software development for computational biology and bioinformatics. Genome Biol 2004; 5: R80.

30. Kauffmann A, Gentleman R, Huber W. ArrayQualityMetrics a bioconductor package for quality assessment of microarray data. Bioinformatics 2009; 25: 415–416.

31. Irizarry RA, Boistad BM, Collin F, Cope LM, Hobbs B et al.Summaries of Affymetrix GeneChip probe level data. Nucleic Acids Res 2003; 31: E15.

32. Smyth GK. Linear models and empirical bayes methods for assessing differential expression in microarray experiments.Stat Appl Genet Mol Biol 2004; 3: 26.

33. Thimm O, Blasing O, Gibon Y, Nagel A, Meyer S et al.MAPMAN: a user-driven tool to display genomics data sets onto diagrams of metabolic pathways and other biological processes. Plant J 2004; 37: 914–939.

34. Usadel B, Nagel A, Steinhauser D, Gibon Y, Blaäsing OE et al. PageMan: an interactive ontology tool to generate, display, and annotate overview graphs for profiling experiments.BMC Bioinformatics 2006; 7: 535.

35. Usadel B, Nagel A, Thimm O, Redestig H, Blaesing OE et al.Extension of the visualization tool MapMan to allow statistical analysis of arrays, display of coresponding genes, and comparison with known responses. Plant Physiol 2005;138: 1195–1204.

36. Deluc LG, Grimplet J, Wheatley MD, Tillett RL, Quilici DR et al. Transcriptomic and metabolite analyses of Cabernet Sauvignon grape berry development. BMC Genomics 2007;8: 429.

37. Pilati S, Perazzolli M, Malossini A, Cestaro A, Dematte L et al. Genome-wide transcriptional analysis of grapevine berry ripening reveals a set of genes similarly modulated during three seasons and occurrence of an oxidative burst at veraison. BMC Genomics 2007; 8: 428.

38. Cheynier V, Duenas-Paton M, Salas E, Maury C, Souquet J-M et al. Structure and properties of wine pigments and tannins. Am J Enol Vitic 2006; 57: 298–305.

39. Pezzuto JM. Grapes and human health. A perspective. J Agric Food Chem 2008; 56: 6777–6784.

40. Tuccio L, Remorini D, Pinelli P, Fierini E, Tonutti P et al.Rapid and non-destructive method to assess in the vineyard grape berry anthocyanins under different seasonal water condition. Austral J Grape Wine Res 2011; 17: 181–189.

41. Zamboni A, Di Carli M, Guzzo F, Stocchero M, Zenoni S et al. Identification of putative stage-specific grapevine berry biomarkers and omics data integration into networks. Plant Physiol 2010; 154: 1439–1459.

42. Martinez-Lüscher J, Sanchez-Diaz M, Delrot S, Aguirreolea J, Pascual I et al. Ultraviolet-B radiation and water deficit interact to alter flavonol and anthocyanin profiles in grapevine berries through transcriptomic regulation. Plant Cell Physiol 2014; 55: 1925–1936.

43. Mattivi F, Guzzon R, Vrhovsek U, Stefanini M, Velasco R.Metabolite profiling of grape: flavonols and anthocyanins. J Agric Food Chem 2006; 54: 7692–7702.

44. Dixon RA, Steele CL. Flavonoids and isoflavonoids – a gold mine for metabolic engineering. Trends Plant Sci 1999; 4: 394–400.

45. Sirtori CR, Lovati MR, Manzoni C, Castiglioni S, Duranti M et al. Proteins of white lupin seed, a naturally isoflavone-poor legume, reduce cholesterolemia in rats and increase LDL receptor activity in HepG2 cells. J Nutr 2004; 134: 18–23.

46. Giribaldi M, Perugini I, Sauvage FX, Schubert A. Analysis of protein changes during grape berry ripening by 2-DE and MALDI-TOF. Proteomics 2007; 7: 3154–3170.

47. Gutierrez-Gonzalez JJ, Guttikonda SK, Tran LSP, Aldrich DL, Zhong R et al. Differential expression expression of isoflavone biosynthetic genes in soybean during water deficits. Plant Cell Physiol 2010; 51: 936–948.

48. Grimplet J, Wheatley MD, Jouira HB, Deluc LG, Cramer GR et al. Proteomic and selected metabolite analysis of grape berry tissues under well-watered and water-deficit stress conditions. Proteomics 2009; 9: 2503–2528.

49. Kennedy JA, Matthews MA, Waterhouse AL. Effect of maturity and vine water status on grape skin and wine flavonoids. Am J Enol Vitic 2002; 53: 268–274.

50. Schijlen EGWM, De Vos CH, Van Tunen AJ, Bovy AG.Modification of flavonoid biosynthesis in crop plants.Phytochemistry 2004; 65: 2631–2648.

51. Downey MO, Dokoozlian NK, Krstic MP. Cultural practice and environmental impacts on the flavonoid composition of grapes and wine: a review of recent research. Am J Enol Vitic2006; 57: 257–268.

52. Moura JCMS, Bonine CAV, de Oliveira Fernandes Viana J, Dornelas MC, Mazzafera P. Abiotic and biotic stresses and changes in the lignin content and composition in plants. J Integr Plant Biol 2010; 52: 360–376.

53. Mitchell WC, Jelenkovic G. Characterizing NAD- and NADP- dependent alcohol dehydrogenase enzymes of strawberries. J Am Soc Hortic Sci 1995; 120: 798–801.

54. Yang L, Wang CC, Guo WD, Li XB, Lu M et al. Differential expression of cell wall related genes in the elongation zone of rice roots under water deficit. Russian J Plant Physiol 2006;53: 390–395.

55. Yoshimura K, Masuda A, Kuwano M, Yokota A, Akashi K.Programmed proteome response for drought avoidance/tolerance in the root of a C_3 xerophyte (wild watermelon) under water deficits. Plant Cell Physiol 2008;49: 226–241.

56. Hoffmann L, Besseau S, Geoffroy P, Ritzenthaler C, Meyer D et al. Silencing of hydroxycinnamoyl-coenzyme A shikimate/quinate hydroxycinnamoyl- transferase affects phenylpropanoid biosynthesis. Plant Cell 2004; 16: 1446–1465.

57. Chowdhury AM, Choi BS, Park SU, Lim HS, Bae H.Transcriptional analysis of hydroxycinnamoyl transferase (HCT) in various tissues of *Hibiscus cannabinus* in response to abiotic stress conditions. Plant Omics J 2012; 5: 305–313.

58. Conde C, Silva P, Fontes N, Dias ACP, Tavares RM et al.Biochemical changes throughout grape berry development and fruit and wine quality. Food 2007; 1: 1–22.

59. Vezzulli S, Civardi SFF, Bavaresco L. Methyl jasmonate treatment as a trigger of resveratrol synthesis in cultivated grapevine. Am J Enol Vitic 2007; 4: 2–5.

60. Deluc LG, Decendit A, Papastamoulis Y, Mérillon JM, Cushman JC et al. Water deficit increases stilbene metabolism in Cabernet Sauvignon berries. J Agric Food Chem 2011; 59: 289–297.

61. Claus H. Laccases: structure, reactions, distribution. Micron2004; 35: 93–96.

62. Pourcel L, Routaboul JM, Cheynier V, Lepiniec L, Debeaujon I. Flavonoid oxidation in plants: from biochemical properties to physiological functions. Trends Plant Sci 2006; 12: 1.

63. Fraignier MP, Marques L, Fleuriet A, Macheix JJ.Biochemical and immunochemical characteristics of polyphenol oxidases from different fruits of Prunus. J Agric Food Chem 1995; 43: 2375–2380.

64. Treutter D. Significance of flavonoids in plant resistance: a review. Environ Chem Lett 2006; 4: 147–157.

65. Liang M, Haroldsen V, Cai X, Wu Y. Expression of a putative laccase gene, ZmLac1, in maize primary roots under stress.Plant Cell Environ 2006; 29: 746–753.

66. Zhang H, Memelink J. Regulation of secondary metabolism by jasmonate hormones. In: Osbourn A E, Lanzotti V, editors.Plant-derived Natural Products: Synthesis, Function and Application. : Springer-Verlag; 2009: pp181–194.

67. Munemasa S, Oda K, Watanabe-Sugimoto M, Nakamura Y, Shimoishi Y et al. The coronatine-insensitive 1 mutation reveals the hormonal signaling interaction between abscisic acid and methyl jasmonate in Arabidopsis guard cells: specific impairment of ion channel activation and second messenger production. Plant Physiol 2007; 143: 1398–1407.

68. Suhita D, Raghavendra AS, Kwak JM, Vavasseur A.Cytoplasmic alkalization precedes reactive oxygen species production during methyl jasmonate and abscisic acid induced stomatal closure. Plant Physiol 2004; 134: 1536–1545.

69. Bell E, Mullet JE. Lipoxygenase gene expression is modulated in plants by water deficit, wounding and methyl jasmonate. Mol Gen Genet 1991; 230: 456–462.

70. Creelman RA, Mullet JE. Biosynthesis and action of jasmonates in plants. Annu Rev Plant Physiol Plant Mol Biol1997; 48: 355–381.

71. Pedranzani H, Sierra-de-Grado R, Vigliocco A, Miersch O, Abdala G. Cold and water stresses produce changes in endogenous jasmonates in two populations of *Pinus pinaster*Ait. Plant Growth Regul 2007; 52: 111–116.

72. Kiribuchi K, Jikumaru Y, Kaku H, Minami E, Hasegawa M et al. Involvement of the basic helix-loop-helix transcription factor RERJ1 in wounding and drought stress responses in rice plants. Biosci Biotechnol Biochem 2005; 69: 1042–1044.

73. Pauwels L, Inze D, Goossens A. Jasmonate-inducible gene: what does it mean? Trends Plant Sci 2009; 14: 87–91.

74. Gadzovska S, Maury S, Delaunay A, Spasenoski M, Joseph C et al. Jasmonic acid elicitation of *Hypericum perforatum* L. cell suspensions and effects on the production of phenylpropanoids and naphtodianthrones. Plant Cell Tissue Organ Cult 2007; 89: 1–13.

75. Shang X, Zhang Y, Peng W, Wang Z, Xie D. Molecular mechanism for jasmonate-induction of anthocyanin accumulation in Arabidopsis. J Exp Bot 2009; 60: 3849–3860.

76. Tassoni A, Fornale S, Franceschetti M, Musiani F, Michael AJ et al. Jasmonates, Na-orthovanadate promote resveratrol production in *Vitis vinifera* cv. Barbera cell cultures. New Phytol 2005; 166: 895–905.

77. Seo S, Sano H, Ohashi Y. Jasmonic acid in wound signal transduction pathways. Physiol Plant 1997; 101: 740–745.

78. Derckel JP, Baillieul F, Manteau S, Audran JC, Haye B et al.Differential induction of grapevine defenses by two strains of*Botrytis cinerea*. Phytopathology 1999; 89: 197–203.

79. Giannakis C, Bucheli CS, Skene KGM, Robinson SP, Scott NS. Chitinase and b-1, 3-glucanase in grapevine leaves: a possible defense against powdery mildew infection. Austral J Grape Wine Res 1998; 4: 14–22.

80. Ferreira RB, Monteiro SS, Piçarra-Pereira MA, Teixeira AR.Engineering grapevine for increased resistance to fungal pathogens without compromising wine stability. Trends Biotechnol 2004; 22: 168–173.

81. Waters EJ, Hayasaka Y, Tattersall DB, Adams KS, Williams PJ. Sequence analysis of grape (*Vitis vinifera*) berry chitinases that cause haze formation in wines. J Agric Food Chem 1998; 46: 4950–4957.

82. Van Loon LC, Rep M, Pieterse CM. Significance of inducible defense-related proteins in infected plants. Annu Rev Phytopathol 2006; 44: 135–162.

83. Park CJ, Kim KJ, Shin R, Park JM, Shin YC et al.Pathogenesis-related protein 10 isolated from hot pepper functions as a ribonuclease in an antiviral pathway. Plant J2004; 37: 186–198.

84. Naidoo R, Myburg A, Berger D, Naidoo S. Expression profiling of putative *Eucalyptus grandis* defence marker genes in response to treatment with methyl jasmonate and salicylic acid. BMC Proc 2011; 5: 96.

85. Liu WX, Zhang FC, Zhang WZ, Song LF, Wu WH et al.*Arabidopsis* Di19 functions as a transcription factor and modulates *PR1*, *PR2*, and *PR5* expression in response to drought stress. Mol Plant 2013; 6: 1487–1502.

86. Vanholme R, Morreel K, Ralph J, Boerjan W. Lignin engineering. Curr Opin Plant Biol 2008; 11: 1–8.

87. Vogt T. Phenylpropanoid biosynthesis. Mol Plant 2010; 3: 2–20.

88. Kutchan TM. Alkaloid biosynthesis: the basis for metabolic engineering of medicinal plants. Plant Cell 1995; 7: 1059–1070.

89. Kaminaga Y, Schnepp J, Peel G, Kish CM, Ben-Nissan G et al. Plant phenylacetaldehyde synthase is a bifunctional homotetrameric enzyme that catalyzes phenylalanine decarboxylation and oxidation. J Biol Chem 2006; 281: 23357–23366.

90. Rippert P, Matringe M. Molecular and biochemical characterization of an *Arabidopsis thaliana* arogenate dehydrogenase with two highly similar and active protein domains. Plant Mol Biol 2002; 48: 361–368.

91. Hiremath PJ, Farmer A, Canon SB, Woodward J, Kudapa H et al. Large-scale transcriptome analysis in chickpea (*Cicer arietinum* L.), an orphan legume crop of the semi-arid tropics of Asia and Africa. Plant Biotechnol J 2011; 9: 922–931.

92. Good AG, Zaplachinski ST. The effects of drought stress on free amino acid accumulation and protein synthesis in*Brassica napus*. Physiol Plantarum 1994; 90: 9–14.

93. Gzik A. Accumulation of proline and pattern of alpha-amino acids in sugar beet plants in response to osmotic, water and salt stress. Environ Exp Bot 1996; 36: 29–38.

94. Song JT, Lu H, McDowell JM, Greenberg JT. A key role for ALD1 in activation of local and systemic defenses in Arabidopsis. Plant J 2004; 40: 200–212.

95. Alcázar R, Altabella T, Marco F, Bortolotti C, Reymond M et al. Polyamines: molecules with regulatory functions in plant abiotic stress tolerance. Planta 2010; 231: 1237–1249.

96. Gill SS, Tuteja N. Polyamines and abiotic stress tolerance in plants. Plant Signal Behav 2010; 5: 26–33.

97. Agudelo-Romero P, Erban A, Sousa L, Pais MS, Kopka J et al. Search for transcriptional and metabolic markers of grape pre-ripening and ripening and insights into specific aroma development in three Portuguese cultivars. PLoS One 2013;8: e60422.

98. Majumdar R, Shao L, Minocha R, Long S, Minocha SC.Ornithine: the overlooked molecule in the regulation of polyamine metabolism. Plant Cell Physiol 2013; 54: 990–1004.

99. Fischer WN, Kwart M, Hummel S, Frommer WB. Substrate specificity and expression profile of amino acid transporters (AAPs) in Arabidopsis. J Biol Chemistry 1995; 270: 16315–16320.

100. Waditee R, Hibino T, Tanaka Y, Nakamura T, Incharoensakdi A et al. Functional characterization of betaine/proline transporters in betaine-accumulating mangrove. J Biol Chem2002; 277: 18373–18382.

101. Cramer GR, Urano K, Delrot S, Pezzotti M, Shinozaki K.Effects of abiotic stress on plants: a systems biology perspective. BMC Plant Biol 2011; 11: 163.

102. Wang M, Vannozzi A, Wang G, Liang YH, Tornielli GB et al.Genome and transcriptome analysis of the grapevine (*Vitis vinifera* L.) WRKY gene family. Hort Res 2014; 1: 14016.

Chapter 11

Proteome changes in banana fruit peel tissue in response to ethylene and high-temperature treatments

Lina Du[1], Jun Song[2], Charles Forney[2], Leslie Campbell Palmer[2], Sherry Fillmore[2] & ZhaoQi Zhang[1]

[1]College of Horticulture, South China Agricultural University, GuangZhou, China

[2]Agriculture and Agri-Food Canada, Atlantic Food and Horticulture Research Centre, 32 Main St., Kentville, Nova Scotia B4N 1J5, Canada

ABSTRACT

Banana (*Musa* AAA group) is one of the most consumed fruits in the world due to its flavor and nutritional value. As a typical climacteric fruit, banana responds to ethylene treatment, which induces rapid changes of color, flavor (aroma and taste), sweetness and nutritional composition. It has also been reported that ripening bananas at temperatures above 24 °C inhibits chlorophyll breakdown and color formation but increases the rate of senescence. To gain fundamental knowledge about the effects of high temperature and ethylene on banana ripening, a quantitative proteomic study employing multiplex peptide stable isotope dimethyl labeling was conducted. In this study, green (immature) untreated banana fruit were subjected to treatment with 10 µL L^{-1} of ethylene for 24 h. After ethylene treatment, treated and untreated fruit were stored at 20 or 30 °C for 24 h.

Fruit peel tissues were then sampled after 0 and 1 day of storage, and peel color and chlorophyll fluorescence were evaluated. Quantitative proteomic analysis was conducted on the fruit peels after 1 day of storage. In total, 413 common proteins were identified and quantified from two biological replicates. Among these proteins, 91 changed significantly in response to ethylene and high-temperature treatments. Cluster analysis on these 91 proteins identified 7 groups of changed proteins. Ethylene treatment and storage at 20 °C induced 40 proteins that are correlated with pathogen resistance, cell wall metabolism, ethylene biosynthesis, allergens and ribosomal proteins, and it repressed 36 proteins that are associated with fatty acid and lipid metabolism, redox–oxidative responses, and protein biosynthesis and modification. Ethylene treatment and storage at 30 °C induced 32 proteins, which were mainly similar to those in group 1 but also included 8 proteins in group 3 (identified as chitinase, cinnamyl alcohol dehydrogenase 1, cysteine synthase, villin-2, leucine-transfer RNA ligase, CP47 protein and calmodulin) and repressed 43 proteins in 4 groups (groups 4–7), of which 6 were associated with photosynthesis II oxygen-evolving protein, the photosynthesis I reaction center, sugar metabolism, the redox–oxidative system and fatty acid metabolism. Differences in the response to ethylene and holding temperature at 30 °C were also revealed and have been discussed. The identities and quantities of the proteins found were linked with quality changes. This study demonstrates that ethylene and high temperature influence banana fruit ripening and senescence at the proteomic level and reveals the mechanisms by which high temperature accelerates banana fruit ripening.

1. INTRODUCTION

Banana (*Musa acuminate*, AAA group) is among the most popular fruits worldwide due to its attractive color, flavor, sweetness and texture. The banana is a typical climacteric fruit that produces a significant burst of autocatalytic ethylene and respiration at the onset of ripening, followed by changes in color, firmness and flavor.[1] Banana fruit are typically harvested at the green mature stage and shipped to wholesale markets, where they are then treated with ethylene to stimulate fruit ripening; this treatment results in a rapid color change from green to yellow and the development of a 'fruity' flavor when stored at 16–24 °C.[2] However, the normal de-greening process can be interrupted if the fruit are stored at temperatures above 24 °C. Under this condition, fruit soften normally but remain green, yielding 'green ripe' bananas.[3,4,5] This response of banana fruit to high post-harvest temperatures, which results in the failure of proper peel color to develop, causes significant economic loss for the banana industry.

Chlorophyll degradation commonly occurs during leaf senescence and fruit ripening in plants. Since the 1990s, research in understanding chlorophyll degradation in leaves has advanced greatly.[6,7,8]Chlorophyll degradation involves a group of chloroplast enzymes including chlorophyll *b* reductase, 7-hydroxymethyl chlorophyll *a* reductase and magnesium dechelatase.[9,10,11] During this degradation, chlorophyll *a* loses Mg and forms pheophorbide *a,* which is further catabolized by pheophorbide *a*oxygenase to form red reducing fluorescence compounds termed fluorescent chlorophyll catabolites (FCCs). Pheophorbide *a* oxygenase further oxidizes FCCs to red chlorophyll catabolites. Primary FCCs are exported from the chloroplast to the cytosol where they can be modified enzymatically. Most FCCs are imported into the vacuole where they are isomerized to colorless compounds.8Research indicates that chlorophyll breakdown in ripening fruit, such as apple and pear, is similar to that in senescent leaves. However, banana fruit present a different chlorophyll degradation pathway, in which large amounts of FCCs with blur luminescence accumulate in the outer region of the peel.12This finding has been recognized as a critical step in chlorophyll degradation.8 'Green ripe' bananas, which occur when bananas are held at temperatures >24 °C, is recognized as a physiological disorder by the banana industry.[2,3] However, in another family of banana (*Musa*, ABB group), normal chlorophyll degradation occurs at 35 °C.13 Other fruit, such as tomato14 and 'Shamouti' orange,15also demonstrate normal chlorophyll degradation at 35 °C. Therefore, the mechanism by which temperature alters the degradation of chlorophyll in banana is unknown. Banana fruit also express rapid senescence and high ethylene production when held at high temperatures; however, exogenous ethylene treatment does not accelerate chlorophyll degradation.2 A comparative study on banana fruit that were stored at 20 or 30 °C revealed that the conversion of starch to sugar is significantly enhanced at 30 °C.16 Similarly, the rapid accumulation of soluble solids has also been associated with 'green ripe' bananas.4 When the gene expression associated with chlorophyll degradation in bananas held at 20 and 30 °C was compared, the senescence of fruit held at 30 °C was accelerated but the expression of genes related to chlorophyll degradation was inhibited.4

As a typical climacteric fruit, banana shows not only a climacteric change in ethylene production and respiration but also a dose response to ethylene treatment; for this reason, banana has been recognized as a model fruit for post-harvest physiology studies.[17,18] Significant research has been reported at both the metabolomic and genomic levels that show the effects of fruit ripening and ethylene treatment on fruit quality indices of banana fruit, including, color,[19] firmness, cell wall composition[20] and aroma/flavor production.[21,22] Ripening-related gene expression in banana peel was examined and compared with that from fruits that were ripened at 20 and 35 °C. A putative ATP-citrate lyase clone was isolated from ripening peel and was downregulated at 35 °C.[19]

At the proteomic level, 50 proteins using two-dimensional gels that respond to fruit ripening and ethylene signal transduction have been identified.[23,24] Another study identified class I chitinase, lipid transfer protein, thaumatin-like protein and 1,3-glucanase as a group of allergens in banana.[25]However, more comprehensive studies employing quantitative proteomic techniques on banana fruit have not been reported. In this study, a quantitative proteomic method employing multiplex peptide stable isotope dimethyl labeling was applied to reveal changes in protein levels in response to ethylene and storage temperature to further investigate the mechanism of the 'green ripe' phenomena at the proteomic level.

2. MATERIALS AND METHODS

2.1 Fruit samples and treatments

Mature green banana fruit (*Musa* spp. AAA group, Cavendish subgroup) that had not been treated with ethylene were purchased from a local wholesale market. For each biological replicate, 20 kg of banana fruit (7–9 hands) were obtained. From each hand, five fruit were randomly selected. The fruit samples were prepared and treated as described previously.[26] Briefly, the fruit were divided into four treatments. Two groups were placed in 15-L glass chambers and exposed to a 0.42 mL s^{-1} flow of 10 μL L^{-1} ethylene for 24 h at 20 °C to initiate ripening. The other two groups were placed in a chamber with air and no ethylene at 20 °C for 24 h. After each treatment, one group of fruit from each treatment was allowed to ripen at 20 °C in unsealed polyethylene plastic bags for 1 day, and the remaining two groups were ripened similarly at 30 °C. Five individual fruit were picked from each group, and fruit color and chlorophyll fluorescence (CF) were measured at three different locations on the peel from bottom to top. CF was measured using an OS-500 modulated Fluorometer (OPTI Sciences, Hudson, NH, USA) operating in the 'Fv/Fm' mode. Fo (minimum fluorescence), Fv (variable fluorescence), Fm (maximum fluorescence) and Fv/Fm (exciton transfer efficiency) were determined after dark-adaptation for 30 min at room temperature.26 Peel tissues were then sampled, separately sliced, mixed and frozen using liquid N_2. The frozen tissue was then pulverized and stored at −85 °C until further use.

2.2 Protein extraction, quantitation and digestion

Proteins were extracted using a modified phenol extraction followed by ammonium acetate–methanol precipitation.27 Proteins were quantitated using the RC/DC protein assay kit (Bio-Rad Laboratories, Hercules, CA, USA); bovine serum albumin was used as the standard.28 Detailed procedures for protein digestion and peptide desalting have been described.29,30 Briefly, the protein samples were diluted with triethylammonium bicarbonate to a final concentration of less than 1 Murea, reduced with dithiothreitol for 30 min at 30 °C and then alkylated with iodoacetamide for 60 min at room temperature in the dark. The proteins were digested with sequencing grade-modified trypsin (Promega, Madison, WI, USA) overnight at 37 °C (final trypsin to substrate ratio of 1:50 (w/w). The peptides were desalted using Oasis HLB cartridges (Waters, Milford, MA, USA). The eluted peptides were then evaporated to dryness using a vacuum centrifuge and re-dissolved in 2% (v/v) ACN and 1% (v/v) formic acid.

2.3 Protein identification and quantitation using stable isotope dimethyl labeling

The resulting peptides of the total tryptic digestion of the various biological samples were labeled with isotopomeric dimethyl labels as described by Boersema *et al.*31 with minor modifications.32 Due to the labeling regime, only three treatment samples could be compared. Bananas that had not been treated with ethylene but had been stored for 2 days at 20 °C were used as the control, while the test samples were ethylene-treated bananas that had been stored for 1 day at 20 °C and 30 °C, respectively. Four hundred (400) µg of protein obtained from samples of each treatment group were re-suspended in 200 µL of IPG buffer comprising 5% glycerol and ampholytes at pH 3.0–10.0 (GE Healthcare, Bio-Sciences AB, Rapsgatan 7E, Uppsala, Sweden). Samples of fruit, including control fruit without ethylene treatment at 20 °C for 2 days, were labeled as 'light' (L), samples obtained after 1 day of ethylene treatment at 20 °C were labeled as 'intermediate' (M) and samples obtained after 1 day of ethylene treatment at 30 °C were labeled as heavy' (H), following incubation with CH_2O (light formaldehyde), CD_2O (intermediate formaldehyde) and $^{13}CD_2O$ (heavy formaldehyde), respectively. Two labeled groups were created from two biological replicates. After labeling, three differentially labeled samples were pooled at a ratio of 1:1:1 (w/w) and desalted prior to OFFGEL peptide fractionation (Supplementary Figure S1).

2.4 OFFGEL fractionation

Total labeled peptides were fractionated into 24 fractions based on *pI* using an OFFGEL-Fractionator (3100, Agilent Technologies, Palo Alto, CA, USA) according to the manufacturer's protocol and as described previously.30

2.5 LC/MS analysis; Q-TOF

Details of the LC/MS procedure using q-TOF have been described previously.32 Briefly, mass spectra were acquired on a Waters Quadrupole Time-of-flight (QTof) liquid chromatography/mass spectrometry (LC/MS) instrument (XeVo, Milford, MA, USA), which was equipped with a nano-electrospray source and a nano-lockspray interface. A lock mass solution of [Glu[1]]-Fibrinopeptide B (200 fmol μL^{-1}) was infused at a frequency of 60:1 (sample-to-reference ratio). A potential of 2800 V was applied to a 20-µm-inner diametre, tapered electrospray tip (FS360-75-30-N-C12, New Objective, Woburn, MA, USA). The cone voltage was set to 30 kV, and the source temperature used was 100 °C. Chromatographic separations were conducted using a nano-Acquity UPLC system (Waters) equipped with an UPLC peptide BEH C_{18} column (75 µm×10 cm×1.7 µm; Waters) in a data-dependent mode. Separation was carried out using a gradient of 5% B to 30% B over 53 min, then increasing to 90% B over 5 min at a flow rate of 400 nL min^{-1}, where A: 0.1% formic acid in water and B: 0.1% formic acid in acetonitrile.

Raw data files from the 24 OFFGEL fractions for each labeling group were combined, and a single database search was performed. Peptide sequence searches were conducted using MASCOT version 2.5.0 (Matrix Science,

London, UK) to search the NCBI using the following parameters: raw MS/MS data were searched against NCBI *viridiplantae* entries, 2 474 089 sequences, updated on 01 December 2014 (NIH, Bethesda, MD, USA). This updated sequence database also contains 43 900 protein sequences for banana. The MS and MS/MS mass tolerances were both 0.3 Da, and one missed cleavage was allowed. Carboxamidomethyl cysteine and oxidized methionine were set as fixed and variable modifications, respectively. Proteins with significant peptide matches were selected for error tolerant searching. MASCOT searches were based on a significance threshold of$P<0.05$, and the MudPIT scoring system (MASCOT scores $\geqslant 52$) was used. False positive peptide identification rates were calculated using the decoy option provided by MASCOT and were estimated as 1.0%. Automated quantification was performed using MASCOT Distiller version 2.5.1.0 (Matrix Science) to calculate the mean protein ratio and s.d. The raw data were also manually inspected.

2.6 Statistical analysis

Four biological experiments were conducted on quality and postharvest physiology. For the proteomic study, two biological replicates were employed. To determine protein function, the proteomic GO (Gene Ontology) tool was used (http://blast.ncbi.nlm.nih.gov/). Protein clustering and heatmap profiles were produced using EPCLUST software (http://www.bioinf.ebc.ee/EP/EP/EPCLUST/) and ClustVis (http://biit.cs.ut.ee/clustvis/). Protein sequences were also searched against the NCBI database employing BLASTp and cut off values of$E$-value 1×10^{-5}. To characterize the GO function group, an enriched analysis was conducted using common identified proteins as a reference.

3. RESULTS AND DISCUSSION

3.1 Physiological changes after ethylene and high-temperature ripening

Physiological changes in banana fruit in response to ethylene treatment and storage at 20 and 30 °C were evaluated over 7 days as reported in a previous study.26 The 'green ripe' phenomenon was observed 4 days after ethylene treatment and high-temperature storage at 30 °C; the hue angle remained >105° compared with <96° for fruit stored at 20 °C. These differences were not apparent after 1 day of storage at 20 and 30 °C, where hue angle was decreased from 119° to 115° and 114°, respectively. Fruit without ethylene treatment remained green. Ethylene and high-temperature ripening caused significant changes in chloroplast function as measured by peel CF.26 Among the CF parameters, Fv/Fm is an indicator of photochemical efficiency. Without ethylene treatment, no change in Fv/Fm was observed (the value remained at ~0.81). With ethylene treatment, however, Fv/Fm decreased to 0.67 and 0.53 following 1 day of storage at 20 °C and 30 °C, respectively, indicating a loss of photochemical efficiency. Fo increased from 296 to 385 at 30 °C, which has been proposed to reflect the effect of stress on chloroplast function.33

3.2 Protein identification and cluster analysis of proteins that exhibited altered expression levels

Significant changes in protein levels in response to ethylene and 30 °C treatments were revealed through a quantitative proteomic investigation that employed peptide dimethylation; this study simultaneously quantified the proteome of three samples with different biological changes that were subjected to both ethylene and storage at 30 °C. A total of 1067 and 1028 proteins were identified and quantified from two individual biological replicates. Among these proteins, 416 were common to both replicates, and of these, the expression of 91 were changed twofold or more in response to treatment; this value (twofold) was used as a cutoff value to distinguish significant increases and decreases of protein abundance when compared with the control.

Seven significant clusters were identified and categorized for these 91 proteins (Figure 1 and Table 1). Cluster 1 comprised 24 proteins that included: PR-1, PR-4-like, pathogenesis-related protein PRB1-2, pathogenesis-related protein 1, alpha-glucosidase Os06g0675700, aldehyde dehydrogenase family 7 member B4, major allergen, endo-beta-1,3-glucanase, pyrophosphate-energized vacuolar membrane proton pump-like, cell wall invertase, pectate lyase 2, phytohormone-binding protein-like, 1-aminocyclopropane-1-carboxylate oxidase, isoflavone reductase, 60S ribosomal protein L30 and allene oxide synthase 2-like proteins, all of which increased after ethylene treatment and storage at 20 or 30 °C for 1 day. The increase in these proteins was greater at 30 °C than at 20 °C. Cluster 2 comprised 16 proteins that increased after ethylene treatment and 1 day of storage at 20 and 30 °C, but the increase was greater at 20 °C than at 30 °C. Proteins in Cluster 2 included: pectin methyl esterase/invertase inhibitor-like protein; alpha-amylase 3; 4-coumarate-CoA ligase-like 10; bifunctional L-3-cyanoalanine synthase/cysteine synthase 2, mitochondrial; cationic peroxidase SPC4-like; carboxylesterase 15; alpha-glucan water dikinase; allene oxide synthase 2-like; patatin-like protein 2; another ACO; pectate lyase 1; beta-amylase; ubiquitin/ribosomal protein 27a; prohibitin-1; and cysteine desulfurase 1. Cluster 3 included eight proteins, the levels of which decreased following ethylene treatment and 1 day of storage at 20 °C but increased when stored at 30 °C compared with those stored at 20 °C. Proteins in Cluster 3 included: chitinase class I; cinnamyl alcohol dehydrogenase 1; cysteine synthase; villin-2; leucine-transfer RNA ligase; CP47 protein and CaM. Cluster 4 comprised seven proteins, including sucrose-phosphate synthase 1, xyloglucan endotransglucosylase/hydrolase protein 7, serine hydroxymethyltransferase, small HSP, proliferation-associated protein 2G4-like, cytochrome c reductase-processing peptidase subunit II and 17.3 kDa class II HSP. This group of proteins increased following ethylene treatment and 1 day of storage at 20 °C but decreased when stored for 1 day at 30 °C. Clusters 5, 6 and 7 included 36 proteins, the levels of which decreased following ethylene treatment regardless of storage temperature. Proteins in Cluster 6 were slightly more abundant after storage at 30 °C than at 20 °C. These significant changes in protein abundance revealed dynamic changes in protein profiles in response to ethylene and storage temperature.

3.3 Protein changes in relation to metabolic pathways

A previous study that employed a differential gene expression analysis indicated 20 non-redundant families of clones that were significantly changed in banana fruit in response to ethylene treatment; these genes included those encoding enzymes involved in ethylene biosynthesis, respiration, starch metabolism, and cell wall degradation.[19,34] Elevated temperature at 35 °C for 2 and 6 days had little or no effect on the expression of most cDNA clones isolated from banana peel, indicating that the effect of high temperature on chlorophyll catabolism is unlikely to be mediated through a general inhibition of ripening.[19] However, ATP-citrate lyase was found to be down-regulated, and it was postulated that new fatty acid biosynthesis occurs during the conversion of chloroplasts to chromoplasts and ATP-CL might play a role in such events. In this quantitative protein investigation of 413 proteins, 91 significantly changed and were identified and quantified. Proteins were categorized as belonging to 9 biological processes according to the GO annotation (Supplementary Figure S2), representing various biological functions including chloroplast and photosynthesis function, redox and reactive oxygen species (ROS) scavenging, stress response, and cell wall, signal, fatty acid, amino acid, starch and sugar metabolism. These proteins with significantly changed abundance might be related to fruit metabolism and could lead to a better understanding of the physiological changes involved in 'green ripe' and fruit ripening of banana fruit.

3.4 Chloroplast and photosynthesis

The proteins that were present at significantly changed abundance and were involved with chloroplast and photosynthesis function included oxygen-evolving enhancing proteins, proteins associated with the photosystem I reaction center subunit, Rubisco, and CP47. The abundances of most of these proteins were either unchanged or decreased in response to ethylene and storage at 20 and 30 °C (Figure 2a). Only CP47 was increased in abundance in response to storage at 30 °C compared with 20 °C. CP47 is one of the core antenna proteins in the photosynthesis II system and interacts with the oxygen-evolving site of photosynthesis II and manganese-stabilizing protein.35However, most of the proteins that were related to oxygen-evolving enhancing proteins were inhibited by both ethylene and storage at 30 °C. The development of oxygenic photosynthesis brought about many changes that required alterations to existing pigments, the generation of an oxygen evolution complex, and protection against the toxic effects of oxygen by-products. Therefore, a decrease of oxygen-evolving enhancing proteins indicates reduced photosynthetic activity, which might be linked to the reduced photosynthetic capacity observed here by the decrease in CF-related parameters (Fv/Fm) and the increase in Fo.

A previous biochemical analysis reported that when banana peel ripened normally at 20 and 35 °C, chlorophyll breakdown products could be isolated from fruit. The colored breakdown products, chlorophyllide and pheophorbide, were not detected in peel that had been subjected to either of the two temperature treatments.19 However, non-fluorescence chlorophyll catabolites accumulated to higher concentrations at 20 °C than at 35 °C.

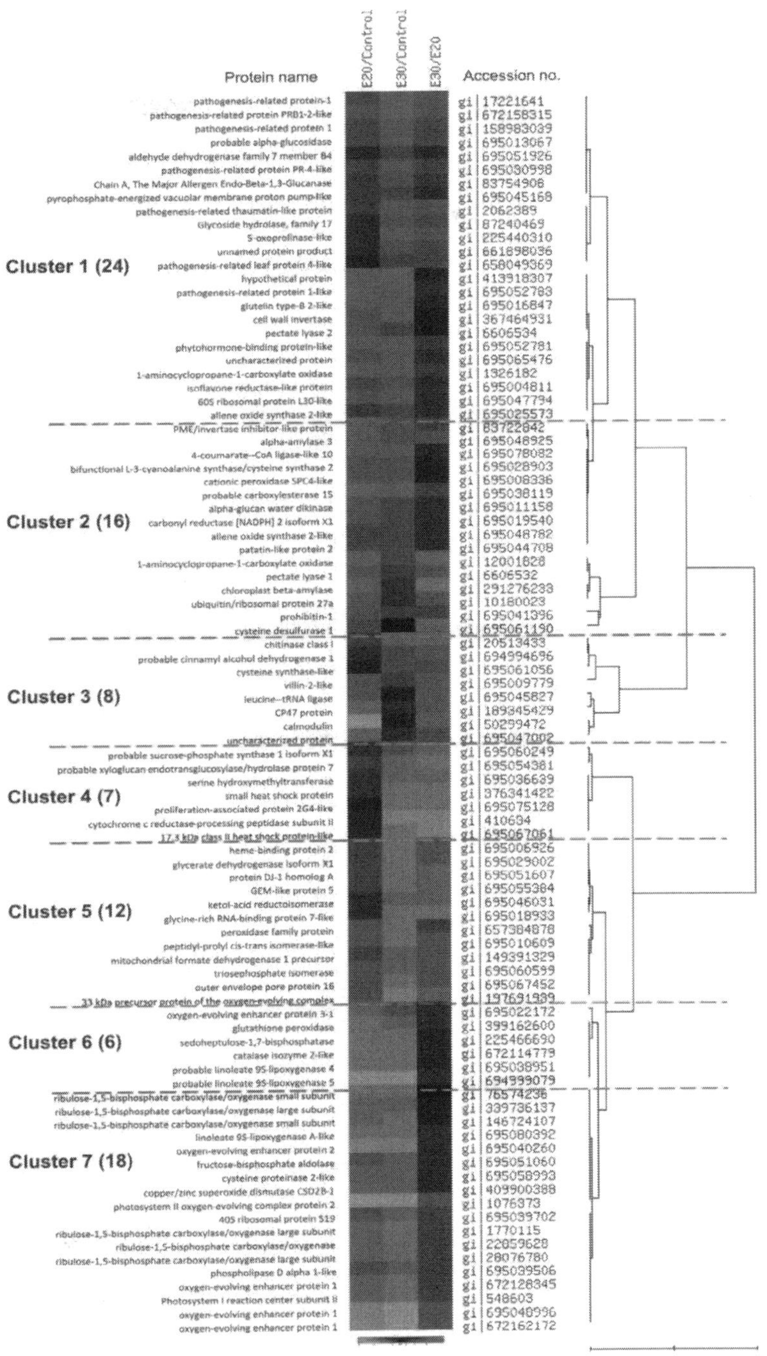

Figure 1. Hierarchical cluster analysis of 91 protein abundances in banana fruit peel tissue after ethylene treatment and stored for 1 day at 20 °C or

30 °C. On the basis of the abundance ratio between ethylene-treated fruit at 20 °C versus control, ethylene-treated fruit at 30 °C versus control as well as ethylene-treated fruit at 30 °C versus ethylene-treated fruit at 20 °C, seven major protein clusters were demonstrated. Cluster 1 includes proteins that were increased in abundance at both 20 °C and 30 °C; Cluster 2 includes proteins that were increased in abundance at both 20 °C and 30 °C, but to a lesser degree at 30 °C; Cluster 3 includes proteins that were decreased in abundance at 20 °C but were present at higher abundance in 30 °C; Cluster 4 includes proteins that were increased in abundance at 20 °C but were present at lower abundance at 30 °C; Cluster 5 includes proteins that were decreased in abundance at both 20 °C and 30 °C and exhibited lower abundance at 30 °C; Cluster 6 includes proteins that were decreased in abundance at both 20 °C and 30 °C and presented at higher abundance at 30 °C; and Cluster 7 includes proteins that were decreased in abundance at both 20 °C and 30 °C but were of similar abundance at both of these temperatures. The protein name is indicated on the right side of the plot and is listed in Table 1. Increasing intensities of red or green color indicate differential up- or down-regulation compared with the control at day 0, respectively. Identified proteins are also marked and listed in Table 1.

3.5 Cell wall metabolism

A group of proteins involved in cell wall metabolism including xyloglucan endotransglucosylase,36 two pectin lyases and pectin methyl esterase were found to increase in ethylene-treated banana after storage at 20 °C. A lesser increase was found in fruit peel that had been stored at 30 °C (Figure 2b). The increase in the abundance of these proteins might explain the significant loss of firmness of 'green ripe' banana fruit after ethylene treatment, a common phenomenon in bananas.4,26 For cassava root during early program cell deterioration, it was reported that the endotransglucosylase hydrolase (XET) and pectin methyl esterase were upregulated, which indicated a remodeling of cell wall components.37 This result is consistent with similar findings for ethylene-treated banana fruit and might be common at senescence.

Table 1. Significantly changed protein abundances in banana peel in response to ethylene treatment and ripening for 1 day at 20 and 30 °C

Cluster 1 (24)											
1	gi|17221641	PR-1 (*Solanum torvum*)	19 196.45	177	10	1.88	0.37	4.79	1.49	2.68	1.32
2	gi|672158315	Pathogenesis-related protein PRB1-2-like (*Phoenix dactylifera*)	17 391.8	182	8	2.25	0.45	6.05	3.53	2.58	1.05
3	gi|158983039	Pathogenesis-related protein 1 (*Musa acuminata*)	17 696.82	1086	27	2.00	1.02	4.05	0.99	2.18	0.62
4	gi|695013067	Probable alpha-glucosidase (*Musa acuminata* subsp. malaccensis)	99 173.86	207	8	1.93	0.58	3.84	0.32	2.05	0.45
5	gi|695051926	Aldehyde dehydrogenase family 7 member B4 (*Musa acuminata* subsp. malaccensis)	55 197.62	134	4	1.41	0.24	2.01	0.46	1.42	0.08
6	gi|695030998	Pathogenesis-related protein PR-4-like (*Musa acuminata* subsp. malaccensis)	16 044.13	941	23	2.08	0.26	3.58	1.00	1.76	0.70
7	gi|83754908	β-1,3-Glucanase (*Musa acuminata*)	33 460.41	6634	153	1.70	0.03	2.55	0.55	1.49	0.30
8	gi|695045168	Pyrophosphate-energized vacuolar membrane proton pump-like (*Musa acuminata* subsp. malaccensis)	80 723.42	603	17	1.44	0.76	2.07	1.65	1.32	0.45
9	gi|2062389	Pathogenesis-related thaumatin-like protein (Oryza Sativa)	25 675.46	540	11	1.50	0.04	3.87	1.19	2.59	0.87
10	gi|87240469	Glycoside hydrolase, family 17 (Medicago truncatula)	40 402.13	90	5	1.31	0.08	2.55	1.37	1.92	0.93
11	gi|225440310	5-Oxoprolinase-like (Vitis vinifera)	139 321.6	104	4	1.28	0.05	2.95	1.97	2.34	1.63
12	gi|661898036	Unnamed protein product (Coffea canephora)	25 049.74	177	6	1.26	0.13	2.47	1.02	2.01	1.01
13	gi|658049369	Pathogenesis-related leaf protein 4-like (Malus domestica)	20 406.33	170	10	1.18	1.17	2.12	1.91	1.96	0.33
14	gi|413918307	Hypothetical protein (Zea mays)	26 204.45	108	2	5.26	4.54	6.45	6.16	1.15	0.18
15	gi|695052783	Pathogenesis-related protein 1-like (*Musa acuminata* subsp. malaccensis)	17 679.15	593	32	4.17	0.15	5.11	0.85	1.22	0.16
16	gi|695016847	Glutelin type-B 2-like (*Musa acuminata* subsp. malaccensis)	38 469.36	520	43	3.11	0.05	3.67	1.55	1.18	0.48
17	gi|367464931	Cell wall invertase (*Musa acuminata* AAA Group)	66 197.93	139	10	2.86	0.65	2.95	0.31	1.05	0.13
18	gi|6606534	Pectate lyase 2 (*Musa acuminata* AAA Group)	49 927.86	476	20	2.27	0.73	2.05	1.35	1.05	0.93
19	gi|695052781	Phytohormone-binding protein-like (*Musa acuminata* subsp. malaccensis)	20 868.93	885	22	2.99	1.73	4.08	1.33	1.49	0.41
20	gi|695065476	Uncharacterized protein (*Musa acuminata* subsp. malaccensis)	16 951.9	316	6	2.66	0.79	3.75	1.96	1.36	0.34
21	gi|1326182	1-Aminocyclopropane-1-carboxylate oxidase (*Musa acuminata* AAA Group)	36 208.74	306	13	3.58	0.37	5.37	1.72	1.53	0.64
22	gi|695004811	Isoflavone reductase-like protein (*Musa acuminata* subsp. malaccensis)	34 070.46	4426	96	1.97	0.46	2.44	0.48	1.24	0.05
23	gi|695047794	60S Ribosomal protein L30-like (*Musa acuminata* subsp. malaccensis)	12 506.98	209	4	2.23	1.51	2.98	1.76	1.39	0.15
24	gi|695025573	Allene oxide synthase 2-like (*Musa acuminata* subsp. malaccensis)	54 314.66	127	2	1.69	1.46	2.00	1.52	1.27	0.20
Cluster 2 (16)											
25	gi|83722842	Pectin methyl esterase (PME)/invertase inhibitor-like protein (*Musa acuminata*)	20 593.01	336	7	4.67	5.12	2.46	2.01	0.73	0.37
26	gi|695048925	Alpha-amylase 3, chloroplastic (*Musa acuminata* subsp. malaccensis)	10 7165.4	274	22	4.53	1.85	2.63	0.41	0.66	0.36
27	gi|695078082	4-Coumarate--CoA ligase-like 10 (*Musa acuminata* subsp. malaccensis)	55 761.83	177	3	5.01	3.67	4.00	2.01	0.89	0.25
28	gi|695028903	Bifunctional L-3-cyanoalanine synthase/cysteine synthase 2, mitochondrial (*Musa acuminata* subsp. malaccensis)	39 758.09	102	4	2.61	0.44	2.34	0.81	0.93	0.47
29	gi|695008336	Cationic peroxidase SPC4-like (*Musa acuminata* subsp. malaccensis)	38 263.96	67	4	2.73	0.29	2.58	1.06	0.93	0.29
30	gi|695038119	Probable carboxylesterase 15 (*Musa acuminata* subsp. malaccensis)	36 552.01	198	6	6.95	0.16	5.21	0.33	0.75	0.03
31	gi|695011158	Alpha-glucan water dikinase, chloroplastic isoform X1 (*Musa acuminata* subsp. malaccensis)	167 012.3	410	28	2.24	0.46	1.97	0.05	0.90	0.16
32	gi|695019540	Carbonyl reductase (NADPH) 2 isoform X1 (*Musa acuminata* subsp. malaccensis)	29 670.51	140	7	2.27	0.52	1.96	0.23	0.87	0.10
33	gi|695048782	Allene oxide synthase 2-like (*Musa acuminata* subsp. malaccensis)	53 716.27	112	5	2.17	0.77	1.80	0.26	0.86	0.19
34	gi|695044708	Patatin-like protein 2 (*Musa acuminata* subsp. malaccensis)	45 457.98	104	9	2.10	0.32	1.74	0.49	0.86	0.37
35	gi|12001828	1-Aminocyclopropane-1-carboxylate oxidase (*Musa acuminata* AAA Group)	34 760.01	245	17	5.48	3.40	2.14	0.76	0.54	0.47
36	gi|6606532	Pectate lyase 1 (*Musa acuminata* AAA Group)	45 277.51	440	20	2.66	1.22	1.58	0.08	0.66	0.27
37	gi|291276233	Chloroplast beta-amylase (*Musa acuminata* AAA Group)	58 123.58	165	7	4.94	4.33	1.40	0.31	0.42	0.30
38	gi|10180023	Ubiquitin/ribosomal protein 27a (Prunus avium)	17 849.02	195	8	2.38	1.66	1.29	0.13	0.69	0.43
39	gi|695041396	Prohibitin-1, mitochondrial-like isoform X1 (*Musa acuminata* subsp. malaccensis)	32 091.69	138	6	2.55	2.58	0.82	0.14	0.72	0.78
40	gi|695061190	Cysteine desulfurase 1, mitochondrial (*Musa acuminata* subsp. malaccensis)	50 174.36	124	3	2.06	0.11	0.99	0.36	0.48	0.20
Cluster 3 (8)											
41	gi|20513433	Chitinase class I (Taxodium distichum)	10 006.37	472	19	0.73	0.30	5.39	2.95	8.96	7.71
42	gi|694994696	Cinnamyl alcohol dehydrogenase 1 (*Musa acuminata* subsp. malaccensis)	39 306.07	128	5	0.84	0.15	1.91	1.31	2.44	1.99
43	gi|695061056	Cysteine synthase-like (*Musa acuminata* subsp. malaccensis)	34 047.46	235	24	0.92	0.35	2.01	1.54	2.01	0.91
44	gi|695009779	Villin-2-like (*Musa acuminata* subsp. malaccensis)	105 919.8	196	4	0.68	0.31	1.43	0.31	2.22	0.57
45	gi|695045827	Leucine--tRNA ligase, cytoplasmic (*Musa acuminata* subsp. malaccensis)	125 534.8	84	4	0.67	0.41	1.07	0.61	2.29	2.30
46	gi|189345429	CP47 protein, partial (chloroplast) (Asplenium viride)	54 629.05	422	15	0.53	0.53	0.83	0.10	2.85	2.63
47	gi|50299472	Calmodulin (Salvia miltiorrhiza)	16 777.33	449	11	0.17	0.07	0.98	0.53	5.60	0.93
48	gi|695047002	Uncharacterized protein LOC103993135 (*Musa acuminata* subsp. malaccensis)	25 047.58	252	5	0.49	0.14	0.84	0.58	1.97	1.76

Cluster 4 (7)											
49	gi|695060249	Probable sucrose-phosphate synthase 1 isoform X1 (*Musa acuminata* subsp. malaccensis)	119 201.2	76	4	1.35	0.45	0.65	0.36	0.46	0.11
50	gi|695054381	Probable xyloglucan endotransglucosylase/hydrolase protein 7 (*Musa acuminata* subsp. malaccensis)	32 871.67	358	14	1.55	1.10	0.48	0.05	0.43	0.33
51	gi|695036639	Serine hydroxymethyltransferase, mitochondrial-like (*Musa acuminata* subsp. malaccensis)	57 592.95	185	6	1.22	0.50	0.47	0.19	0.46	0.34
52	gi|376341422	Small heat shock protein (*Musa acuminata* AAA Group)	17 746.54	349	12	1.32	0.50	0.43	0.01	0.35	0.14
53	gi|695075128	Proliferation-associated protein 2G4-like (*Musa acuminata* subsp. malaccensis)	43 804.25	145	7	1.06	0.22	0.43	0.09	0.43	0.17
54	gi|410634	Cytochrome c reductase-processing peptidase subunit II, MPP subunit II, P53 (potatoes, var. Marfona, tuber, Peptide Mitochondrial, 530 aa)	59 463.69	136	4	1.12	1.01	0.28	0.21	0.28	0.06
55	gi|695067061	17.3 kDa class II heat shock protein-like (*Musa acuminata* subsp. malaccensis)	17 626.59	795	23	1.19	0.34	0.35	0.00	0.31	0.09
Cluster 5 (12)											
56	gi|695006926	Heme-binding protein 2 (*Musa acuminata* subsp. malaccensis)	24 202.41	223	9	0.77	0.33	0.44	0.07	0.60	0.16
57	gi|695029002	Glycerate dehydrogenase isoform X1 (*Musa acuminata* subsp. malaccensis)	42 297.46	251	10	0.81	0.19	0.44	0.21	0.59	0.40
58	gi|695051607	Protein DJ-1 homolog A (*Musa acuminata* subsp. malaccensis)	47 213.36	142	3	0.76	0.16	0.38	0.02	0.52	0.14
59	gi|695055384	GEM-like protein 5 (*Musa acuminata* subsp. malaccensis)	28 200.4	73	3	0.80	0.24	0.38	0.38	0.57	0.65
60	gi|695046031	Ketol-acid reductoisomerase, chloroplastic-like (*Musa acuminata* subsp. malaccensis)	63 901	856	39	0.90	0.15	0.50	0.11	0.57	0.22
61	gi|695018933	Glycine-rich RNA-binding protein 7-like (*Musa acuminata* subsp. malaccensis)	16 140.95	225	11	0.89	0.03	0.35	0.22	0.40	0.27
62	gi|657384878	Peroxidase family protein (Medicago truncatula)	38 096.33	84	4	0.68	0.42	0.34	0.28	0.78	0.90
63	gi|695010609	Peptidyl-prolyl cis-trans isomerase-like (*Musa acuminata* subsp. malaccensis)	18 575.62	1718	58	0.62	0.05	0.43	0.15	0.68	0.19
64	gi|149391329	Mitochondrial formate dehydrogenase 1 precursor (Oryza sativa Indica Group)	15 437.19	93	5	0.76	0.20	0.46	0.38	0.70	0.69
65	gi|695060599	Triosephosphate isomerase, chloroplastic-like (*Musa acuminata* subsp. malaccensis)	33 611.81	421	19	0.70	0.15	0.49	0.23	0.68	0.18
66	gi|695067452	Outer envelope pore protein 16, chloroplastic (*Musa acuminata* subsp. malaccensis)	15 497.52	209	8	0.66	0.03	0.41	0.10	0.63	0.18
67	gi|197691939	33 kDa precursor protein of the oxygen-evolving complex (Salicornia europaea)	35 196.39	165	9	0.58	0.03	0.29	0.21	0.51	0.38
Cluster 6 (6)											
68	gi|695022172	Oxygen-evolving enhancer protein 3-1, chloroplastic-like (*Musa acuminata* subsp. malaccensis)	24 720.82	212	14	0.49	0.04	0.65	0.21	1.32	0.33
69	gi|399162600	Glutathione peroxidase (*Musa acuminata* AAA Group)	18 691.94	374	13	0.37	0.16	0.47	0.21	1.27	0.01
70	gi|225466690	Sedoheptulose-1,7-bisphosphatase, chloroplastic (Vitis vinifera)	42 898.17	98	3	0.43	0.26	0.44	0.17	1.12	0.27
71	gi|672114779	Catalase isozyme 2-like (*Phoenix dactylifera*)	56 378.56	1627	36	0.40	0.17	0.41	0.01	1.10	0.43
72	gi|695038951	Probable linoleate 9S-lipoxygenase 4 (*Musa acuminata* subsp. malaccensis)	76 399	253	17	0.40	0.16	0.43	0.02	1.20	0.55
73	gi|694999079	Probable linoleate 9S-lipoxygenase 5 (*Musa acuminata* subsp. malaccensis)	98 781.88	1498	85	0.18	0.03	0.24	0.08	1.29	0.21
Cluster 7 (18)											
74	gi|76574236	Chloroplast ribulose-1,5-bisphosphate carboxylase/oxygenase small subunit (*Musa acuminata* AAA Group)	20 807.84	208	10	0.45	0.29	0.43	0.24	1.00	0.11
75	gi|339736137	Ribulose-1,5-bisphosphate carboxylase/oxygenase large subunit (Ostreobium quekettii)	15 095.4	221	10	0.52	0.21	0.48	0.12	0.96	0.17
76	gi|146724107	Chloroplast ribulose-1,5-bisphosphate carboxylase/oxygenase small subunit (*Musa acuminata* AAA Group)	20 849.92	566	28	0.48	0.24	0.44	0.13	0.97	0.22
77	gi|695080392	Linoleate 9S-lipoxygenase A-like (*Musa acuminata* subsp. malaccensis)	96 773.43	85	6	0.43	0.12	0.38	0.07	0.90	0.09
78	gi|695040260	Oxygen-evolving enhancer protein 2, chloroplastic-like (*Musa acuminata* subsp. malaccensis)	28 178.71	1284	26	0.32	0.20	0.26	0.12	0.87	0.15
79	gi|695051060	Fructose-bisphosphate aldolase, chloroplastic-like (*Musa acuminata* subsp. malaccensis)	42 279.08	313	14	0.58	0.29	0.48	0.12	0.88	0.24
80	gi|695058993	Cysteine proteinase 2-like (*Musa acuminata* subsp. malaccensis)	34 996.01	389	18	0.54	0.17	0.49	0.29	0.86	0.28
81	gi|409900388	Copper/zinc superoxide dismutase CSD2B-1 (*Musa acuminata*)	22 048.99	107	6	0.53	0.09	0.46	0.15	0.85	0.13
82	gi|1076373	Photosystem II oxygen-evolving complex protein 2—Arabidopsis thaliana (fragment)	1434.2	163	8	0.23	0.06	0.17	0.11	0.69	0.30
83	gi|695039702	40S Ribosomal protein S19 (*Musa acuminata* subsp. malaccensis)	16 103.94	273	16	0.62	0.21	0.49	0.30	0.75	0.24
84	gi|1770115	Ribulose-1,5-bisphosphate carboxylase/oxygenase large subunit (Casasia clusiifolia)	52 706.64	1337	51	0.57	0.28	0.38	0.08	0.73	0.22
85	gi|22859628	Ribulose-1,5-bisphosphate carboxylase/oxygenase, partial (chloroplast) (*Selaginella pilifera*)	47 457.51	1470	58	0.53	0.13	0.36	0.04	0.71	0.25
86	gi|28076780	Ribulose-1,5-bisphosphate carboxylase/oxygenase large subunit (*Ulva lactuca*)	50 182.23	1185	40	0.53	0.21	0.36	0.03	0.74	0.24
87	gi|695039506	Phospholipase D alpha 1-like (*Musa acuminata* subsp. malaccensis)	93 215.93	98	3	0.59	0.13	0.48	0.28	0.77	0.31
88	gi|672128345	Oxygen-evolving enhancer protein 1, chloroplastic-like (*Phoenix dactylifera*)	35 656.52	522	12	0.50	0.16	0.39	0.35	0.71	0.47
89	gi|548603	Photosystem I reaction center subunit II (Hordeum vulgare)	21 976.88	501	20	0.54	0.07	0.43	0.02	0.79	0.14
90	gi|695048996	Oxygen-evolving enhancer protein 1, chloroplastic-like (*Musa acuminata* subsp. malaccensis)	36 229.49	350	15	0.34	0.16	0.23	0.17	0.63	0.20
91	gi|672162172	Oxygen-evolving enhancer protein 1, chloroplastic (*Phoenix dactylifera*)	35 547.58	465	11	0.33	0.10	0.22	0.17	0.62	0.31

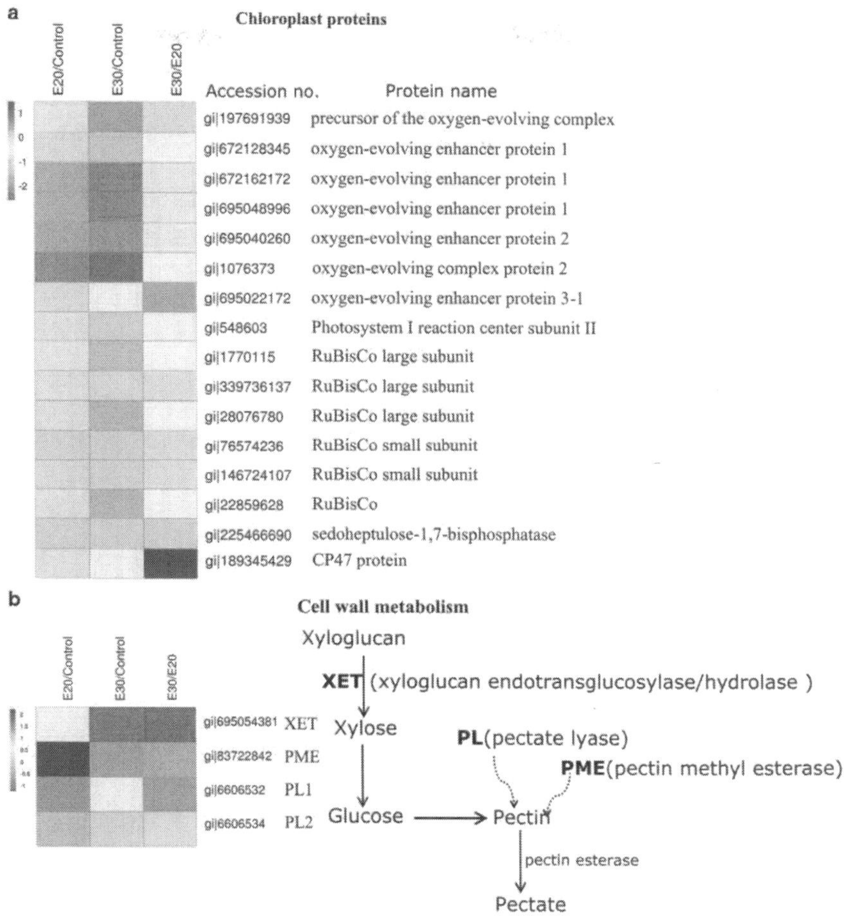

Figure 2. Proteins with significantly changed abundances as classified into putative functional group category—(**a**) chloroplast proteins; (**b**) cell wall metabolism.

3.6 ROS scavenging and stress response

With ROS scavenging proteins, a significant decrease of catalase (CAT), superoxide dismutase and glutathione peroxidase was observed in ethylene-treated fruit held at both 20 and 30 °C (Figure 3a). In contrast, the abundance of cationic peroxidase was increased at both temperatures, whereas 5-oxoprolinase was more abundant in fruit held at 30 °C (Figure 3a). The decreases in the abundances of CAT, superoxide dismutase and glutathione peroxidase indicated a possible down-regulation of ROS scavenging capacity under these treatments. Proteomic studies in apples showed that superoxide dismutase and glutathione peroxidase increased with fruit ripening and in response to ethylene treatment, but CAT tended to decrease.29 In cassava root, both superoxide dismutase and CAT increased during post-harvest physiological deterioration. The increases in the abundances of cationic peroxidase and 5-oxoprolinase in fruit held at 30 °C indicated that ROS scavenging and glutathione (GSH) glutamate metabolism are linked. In Arabidopsis, the role of 5-oxoprolinase in GSH degradation to glutamate has been reported.38 One of the roles of GSH degradation is to provide Cys for protein synthesis and as a precursor for numerous metabolites; this might explain why cysteine synthase was increased in conjunction with ethylene and storage at 30 °C. The result obtained implies that GSH degradation in banana peel forms part of the response to ethylene and 30 °C storage.

A group of stress response proteins was also found to increase in abundance after treatment with ethylene and storage at 30 °C (Figure 3b). These proteins include various pathogenesis-related proteins, phytohormone-binding protein, β-1,3-glucanase, isoflavone reductase and chitinase class I protein. Significant increases in the abundances of pathogenesis-related protein, β-1,3-glucanase and chitinase I proteins have been reported in gene expression34 and proteomic studies,24 indicating that these proteins respond to both fruit ripening and 30 °C storage. In contrast, the abundances of heat shock proteins (HSP) and glycine-rich RNA-binding protein decreased in response to ethylene and holding at 30 °C. Formate dehydrogenase, phospholipidase and cysteine proteinase 2 decreased in response to ethylene treatment, but showed only little change with holding at 30 °C compared with 20 °C, demonstrating a possible downregulation of these proteins after ethylene treatment.

3.7 Signaling pathway

In terms of signaling pathways, two ACOs and AOS increased in response to ethylene treatment and storage at 30 °C (Figure 4a), whereas three LOX proteins (LOX4, LOX5 and LOXA) and CaM decreased. LOX proteins have been found to be important in fruit ripening and flavor formation in apples39,40 and in other fruit, including tomato41 and peach.42 In banana fruit, differential expression of LOX is observed between peel and pulp tissue after ethylene treatment, with an overall decrease in abundance in the peel.26 The proteomic data obtained here confirm the previously reported *LOX*gene expression data. CaM has an important role in plant stress responses, and its gene expression is often induced by various stresses.43 CaM expression has an important signal transduction role during heat stress, and several HSPs are expressed in correlation with the accumulation of CaM transcripts and proteins in plants.44 However, we found that one small HSP and one 17.3 kDa class II HSP decreased in banana peel after ethylene treatment and during ripening at 30 °C (Figure 3b). It has been reported that seven genes encoding calcium/CaM-regulated proteins respond differently to various stress signals (cold, wounding, pathogen infection, salicylic acid and methyl jasmonate treatment) in mature green tomato fruit. Among these genes, only one exhibited increased expression in response to all treatments, a finding that indicated that individual genes in this Solanum lycopersicum stress responsive (SlSRs) has a distinct role in response to stress and might act as a coordinator between calcium and other stress signals.45 Little is known about CaM in banana fruit, and the decrease of CaM abundance in response to ethylene warrants further investigation.

3.8 Amino acid metabolism and volatile biosynthesis

Proteins involved with amino acid metabolism were also found to increase in response to ethylene treatment and storage at 30 °C. These proteins included cysteine synthase, cysteine desulfurase, serine hydroxyl-methyltransferase, glycerate dehydrogenase isomer, keto-acid reductoisomerase and leucine–transfer RNA ligase (Figure 4b). Among these proteins, one of the cysteine synthases 2 increased in response to both ethylene and storage at 30 °C; another cysteine synthase increased in abundance only in response to holding at 30 °C. Cysteine desulfurase increased in response to ethylene but not in response to storage at 30 °C. Cysteine and related compounds not only have important roles in plant metabolism (including the synthesis of defense compounds) but also act as signals for regulating essential plant processes, such as the redox signaling process, photosynthesis, plant protection, plant immunity and root development; this signaling occurs in various cellular compartments.46,47

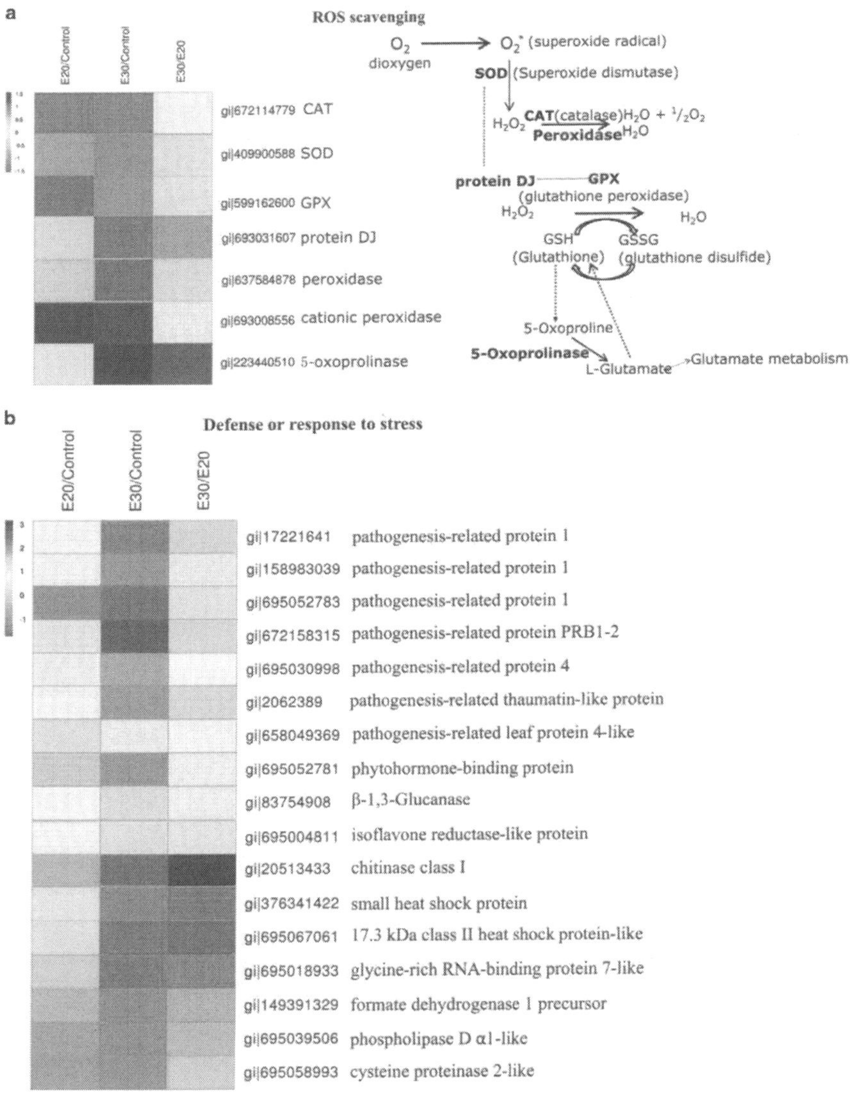

Figure 3. Proteins with significantly changed abundances as classified into putative functional group category—(**a**) ROS scavenging; (**b**) defense or response to stress.

Cysteine desulfurase is a pyridoxal 5'-phosphate-dependent enzyme that catalyzes the conversion of L-cysteine to L-alanine and sulfane sulfur. This group of proteins participates in various biological processes including electron transfer, the regulation of gene expression, iron–sulfur storage photosynthesis and other enzyme activities.[48,49] The identified proteins Cys synthase 2 and Cys desulfurase are located in the mitochondria. Cys synthase 2 acts as a L-3-cyanoalanine synthase, which catalyses the production of cyanide in response to ACO increase as a form of detoxification. Another cysteine synthase protein responded to high temperature but not to ethylene. This result confirms the association between increased levels of 5-oxoprolinase (responsible for GSH degradation) with glutamate, resulting in a possible increase of Cys as mentioned in the previous section describing ROS. This indicates that a complex system is used to homeostatically maintain Cys for normal plant function and in response to both biotic and abiotic stresses. The increases in Cys synthase and Cys desulfurase abundances in response to ethylene treatment might indicate the need to balance Cys levels in banana peel and to compensate for the Cys increase that occurs in response to storage at 30 °C. Our proteomic study on changes in the abundance of Cys-related proteins in response to ethylene treatment and storage at 30 °C sheds light on the importance of these groups of proteins in plant metabolism. However, further investigation on cysteine metabolism during banana fruit ripening and in response to various stresses is warranted.

Previous studies revealed changes in the gene expressions of alcohol acyl transferase, aldehyde dehydrogenase, pyruvate dehydrogenase, hydroperoxide lyase (HPL), lipoxygenase[50] and branched-chain amino acid aminotransferase (BCAT), which have important roles in the formation of volatiles;[26] these enzymes were increased during fruit ripening, especially in response to ethylene treatment. Our proteome study found significant increases in aldehyde dehydrogenase family 7 proteins, which catalyze the conversion between aldehydes and alcohols.

3.9 Starch and sugar metabolism

Significant changes in proteins involved with starch and sugar metabolism were also found. α- and β-amylases as well as glucan water dikinase increased in banana fruit that were treated with ethylene and held at 30 °C (Figure 5). In contrast, fructose-bisphosphate aldolase and triosephosphate isomerase were decreased in abundance. At 30 °C, however, the abundances of most of these proteins were decreased to a greater extent than they were decreased at 20 °C.

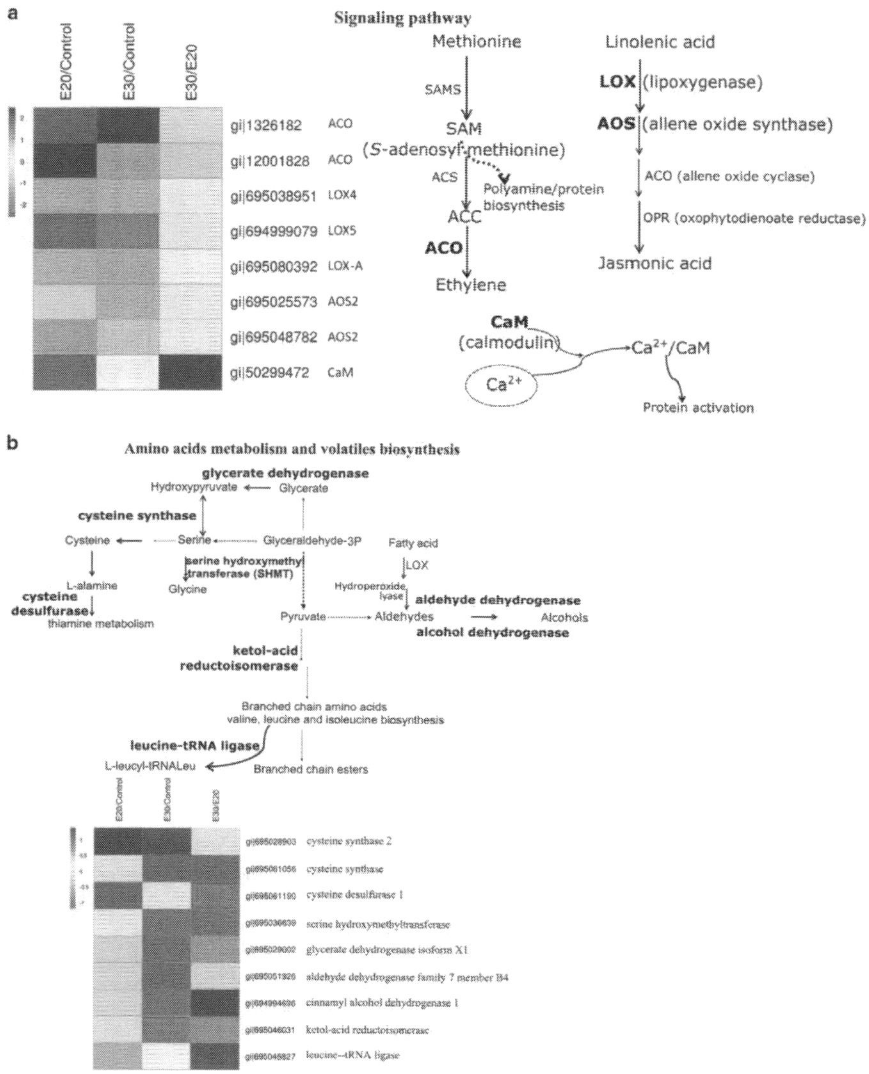

Figure 4. Proteins with significantly changed abundances as classified into putative functional group category—(**a**) signal pathway; (**b**) amino acid metabolism and volatile compound biosynthesis.

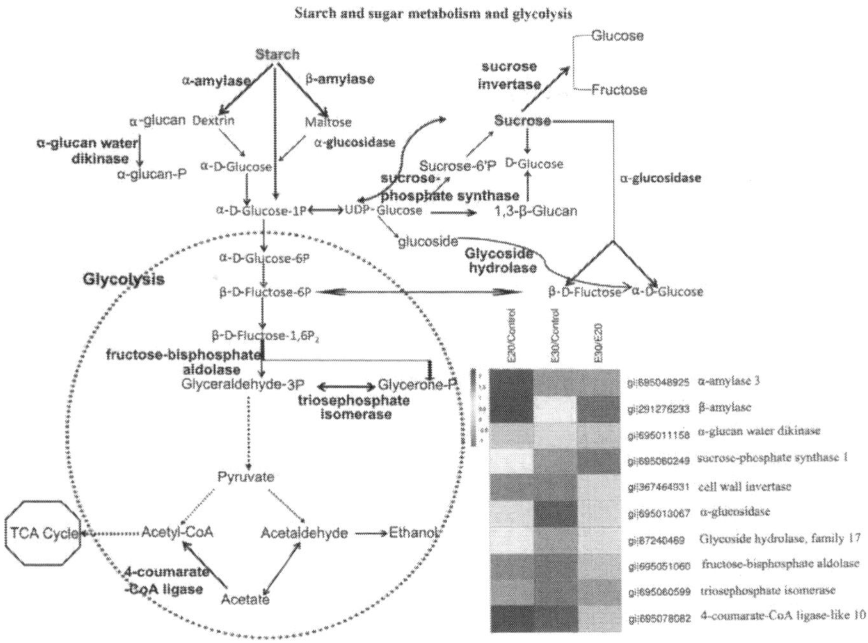

Figure 5. Proteins with significantly changed abundances as classified into putative functional group category—starch and sugar metabolism, as well as glycolysis.

In addition, the abundance of 16 proteins, glutelin type-B 2-like, paratin-like protein 2, prohibitin-1, mitochondrial-like isoform X1, phytohormone-binding protein, and villin-2-like proteins were increased after ethylene treatment and were further increased when held at 30 °C. The abundances of cytochrome c reductase-processing peptidase subunit II, MPP subunit II, P53 protein, outer envelope pore protein 16, 40S ribosomal protein S19, and proliferation-associated protein 2G4 were changed slightly in response to ethylene but were decreased significantly when held at 30 °C compared with 20 °C, indicating these proteins are sensitive to high-temperature storage.

4. CONCLUSIONS

Banana fruit peel tissue showed significant physiological changes in color and CF after ethylene treatment and storage at 20 and 30 °C. Decreases of

Fv/Fm were observed, indicating decreased photosynthetic capacity in peel tissue. Peptide dimethylation in combination with the OFFGEL technique is an effective and high throughput approach that can identify and quantify a large number of proteins and reveal their response to ethylene and storage temperature in banana peel tissue. Significant changes in proteins abundance contribute to the abnormal ripening of banana peel under these conditions. Among the proteins exhibiting significantly changed abundance, multi-pathway proteins such as those involved in chloroplast function, ROS scavenging, signal pathways, defense and the stress response, as well as proteins involved in cell wall, amino acid and sugar metabolism were affected by the treatments. These results revealed at the protein level the molecular metabolism that contributes to abnormal peel color when fruit are held at 30 °C. The identified proteins can be used as targets for further proteomic analysis using MRM to investigate their role, both individually and as a group, in the response of banana fruit to ethylene treatment and high temperature.

ACKNOWLEDGEMENTS

We thank Dr John Delong at Agriculture and Agri-Food, Canada, for his critical review of this manuscript. We thank MOE and AAFC for PhD fellowships provided to Lina Du. This project was funded by the A-Base research (RBPI 197) of Agriculture & Agri-Food, Canada.

REFERENCES

1. Seymour G. Banana. In: Seymour G, Taylor J, Tucker G (eds.). Biochemistry of fruit ripening. Netherlands: Springer, 1993: 83–106.

2. Seymour GB, John P, Thompson AK. Inhibition of degreening in the peel of bananas ripened at tropical temperatures. I. Effect of high temperature on changes in the pulp and peel during ripening.Ann Appl Biol 1987; 110: 153–161.

3. Blackbourn HD, Jeger MJ, John P, Thompson AK. Inhibition of degreening in the peel of bananas ripened at tropical temperatures. III. Changes in plastid ultrastructure and chlorophyll-protein complexes

accompanying ripening in bananas and plantains. Ann Appl Biol 1990; 117: 147–161.

4. Yang X, Zhang Z, Joyce D, Huang X, Xu L, Pang X. Characterization of chlorophyll degradation in banana and plantain during ripening at high temperature. Food Chemistry 2009; 114: 383–390.

5. Plainsirichai M, Turner DW. Storage temperature during the early stage of ripening determines whether or not the peel of banana (Musa spp. AAA, Cavendish subgroup) degreens. Fruits 2010; 65: 69–74.

6. Matile P, Hörtensteiner S, Thomas H. Chlorophyll degradation. Annu Rev Plant Physiol Plant Mol Biol 1999; 50: 67–95.

7. Hörtensteiner S. Chlorophyll degradation during senescence. Ann Rev Plant Biol 2006; 57: 55–57.

8. Müller T, Kräutler B. Chlorophyll breakdown as seen in bananas: Sign of aging and ripening-a mini-review. Gerontology 2011; 57: 521–527.

9. Kusaba M, Ito H, Morita R, Iida S, Sato Y, Fujimoto M et al. Rice non-yellow coloring1 is involved in light-harvesting complex II and grana degradation during leaf senescence. Plant Cell 2007; 19: 1362–1375.

10. Horie Y, Ito H, Kusaba M, Tanaka R, Tanaka A. Participation of chlorophyll b reductase in the initial step of the degradation of light-harvesting chlorophyll a/b-protein complexes in Arabidopsis. J Biol Chem 2009; 284: 17449–17456.

11. Sato Y, Morita R, Katsuma S, Nishimura M, Tanaka A, Kusaba M. Two short-chain dehydrogenase/reductases, NON-YELLOW COLORING 1 and NYC1-LIKE, are required for chlorophyll b and light-harvesting complex II degradation during senescence in rice. Plant J 2009; 57: 120–131.

12. Moser S, Müller T, Holzinger A, Lütz C, Kräutler B. Structures of chlorophyll catabolites in bananas (*Musa acuminata*) reveal a split path of chlorophyll breakdown in a ripening fruit. Chemistry 2012;18: 10873–10885.

13. Yang X, Pang X, Xu L, Fang R, Huang X, Guan P et al. Accumulation of soluble sugars in peel at high temperature leads to stay-green ripe banana fruit. J Exp Botany 2009; 60: 4051–4062.

14. Ogura N, Nakaya H, Takehana H. Studies on the storage temprature of tomato fruit. I. Effect of high temperature -short term storage of mature green tomato fruits on changes of their chemical composition after ripening at room temperature. J Agric Chem Soc Jpn 1975; 49: 189–196.

15. Cohen E. The effect of temperature and relative humidity during degreening on the colouring of Shamouti orange fruit. J Hortic Sci Biotechnol 1978; 53: 143–146.

16. Pang XQ, Yang XT, Zhang ZQ. Chlorophyll degradation and its control in postharvest fruits. Stewart Postharvest Rev 2008; 4: 1–4.

17. Golding JB, Shearer D, McGlasson WB, Wyllie SG. Relationships between respiration, ethylene, and aroma production in ripening banana. J Agric Food Chem 1999; 47: 1646–1651.

18. Manrique-Trujillo SM, Ramírez-López AC, Ibarra-Laclette E, Angel Gómez-Lim M. Identification of genes differentially expressed during ripening of banana. J Plant Physiol 2007; 164: 1037–1050.

19. Drury R, Hörtensteiner S, Donnison I, Bird CR, Seymour GB. Chlorophyll catabolism and gene expression in the peel of ripening banana fruits. Physiol Plant 1999; 107: 32–38.

20. Amnuaysin N, Jones ML, Seraypheap K. Changes in activities and gene expression of enzymes associated with cell wall modification in peels of hot water treated bananas. Sci Hort 2012; 142: 98–104.

21. Beekwilder J, Alvarez-Huerta M, Neef E, Verstappen FW, Bouwmeester HJ, Aharoni A. Functional characterization of enzymes forming volatile esters from strawberry and banana. Plant Physiol 2004;135: 1865–1878.

22. Jayanty S, Song J, Rubinstein NM, Chong A, Beaudry RM. Temporal relationship between ester biosynthesis and ripening events in bananas. J Am Soc Hortic Sci 2002; 127: 998–1005.

23. Esteve C, D'Amato A, Marina ML, García MC, Righetti PG. In-depth proteomic analysis of banana (Musa spp.) fruit with combinatorial peptide ligand libraries. Electrophoresis 2013; 34: 207–214.

24. Toledo TT, Nogueira SB, Cordenunsi BR, Gozzo FC, Pilau EJ, Lajolo FM et al. Proteomic analysis of banana fruit reveals proteins that are differentially accumulated during ripening. Postharvest Biology and Technology 2012; 70: 51–58.

25. Palacin A, Quirce S, Sanchez-Monge R, Bobolea I, Diaz-Perales A, Martin-Muñoz F et al.Sensitization profiles to purified plant food allergens among pediatric patients with allergy to banana.Pediatr Allergy Immunol 2011; 22: 186–195.

26. Yang X, Song J, Fillmore S, Pang X, Zhang Z. Effect of high temperature on color, chlorophyll fluorescence and volatile biosynthesis in green-ripe banana fruit. Postharvest Biol Technol 2011; 62: 246–257.

27. Zheng Q, Song J, Doncaster K, Rowland E, Byers DM. Qualitative and quantitative evaluation of protein extraction protocols for apple and strawberry fruit suitable for two-dimensional electrophoresis and mass spectrometry analysis. J Agric Food Chem 2007; 55: 1663–1673.

28. Hustoft HK, Reubsaet L, Greibrokk T, Lundanes E, Malerod H. Critical assessment of accelerating trypsination methods. J Pharm Biomed Anal 2011; 56: 1069–1078.

29. Zheng Q, Song J, Campbell-Palmer L, Thompson K, Li L, Walker B et al. A proteomic investigation of apple fruit during ripening and in response to ethylene treatment. J Proteomics 2013; 93: 276–294.

30. Yang X, Li L, Song J, Palmer LC, Li X, Zhang Z. Peptide prefractionation is essential for proteomic approaches employing multiple-reaction monitoring of fruit proteomic research. J Sep Sci 2014; 37: 77–84.

31. Boersema PJ, Raijmakers R, Lemeer S, Mohammed S, Heck AJR. Multiplex peptide stable isotope dimethyl labeling for quantitative proteomics. Nat Protoc 2009; 4: 484–494.

32. Li L, Song J, Kalt W, Forney C, Tsao R, Pinto D et al. Quantitative proteomic investigation employing stable isotope labeling by peptide dimethylation on proteins of strawberry fruit at different ripening stages. J Proteomics 2013; 94: 219–239.

33. Wright H, DeLong J, Harrison PA, Gunawardena A H L A N, Prange RK. The effect of temperature and other factors on chlorophyll a

fluorescence and the lower oxygen limit in apples (Malus domestica.Postharvest Biol Technol 2010; 55: 21–28.

34. Clendennen SK, May GD. Differential gene expression in ripening banana fruit. Plant Physiol 1997;115: 463–469.

35. Bricker TM, Frankel LK. The structure and function of CP47 and CP43 in Photosystem II.Photosynth Res 2002; 72: 131–146.

36. Shah PK, Perez-Iratxeta C, Bork P, Andrade MA. Information extraction from full text scientific articles: Where are the keywords? BMC Bioinformatics 2003; 4: 20.

37. Owiti J, Grossmann J, Gehrig P, Dessimoz C, Laloi C, Hansen MB et al. ITRAQ-based analysis of changes in the cassava root proteome reveals pathways associated with post-harvest physiological deterioration. Plant J 2011; 67: 145–156.

38. Ohkama-Ohtsu N, Oikawa A, Zhao P, Xiang C, Saito K, Oliver DJ. A γ-glutamyl transpeptidase-independent pathway of glutathione catabolism to glutamate via 5-oxoproline in Arabidopsis. Plant Physiol 2008; 148: 1603–1613.

39. Schiller D, Contreras C, Vogt J, Dunemann F, Defilippi BG, Beaudry R et al. A dual positional specific lipoxygenase functions in the generation of flavor compounds during climacteric ripening of apple. Hort Res 2015; 2: 1–13.

40. Vogt J, Schiller D, Ulrich D, Schwab W, Dunemann F. Identification of lipoxygenase (LOX) genes putatively involved in fruit flavour formation in apple (Malus×domestica). Tree Genet Genomes 2013;9: 1493–1511.

41. Chen G, Hackett R, Walker D, Taylor A, Lin Z, Grierson D. Identification of a specific isoform of tomato lipoxygenase (TomloxC) involved in the generation of fatty acid-derived flavor compounds.Plant Physiol 2004; 136: 2641–2651.

42. Han M, Zhang T, Zhao C, Zhi J. Regulation of the expression of lipoxygenase genes in Prunus persica fruit ripening. Acta Physiol Plant 2011; 33: 1345–1352.

43. Snedden WA, Fromm H. Calmodulin as a versatile calcium signal transducer in plants. New Phytol2001; 151: 35–66.

44. Zhang W, Zhou RG, Gao YJ, Zheng SZ, Xu P, Zhang SQ et al. Molecular and genetic evidence for the key role of AtCaM3 in heat-shock signal transduction in Arabidopsis. Plant Physiol 2009; 149: 1773–1784.

45. Yang T, Peng H, Whitaker BD, Jurick WM. Differential expression of calcium/calmodulin-regulated SlSRs in response to abiotic and biotic stresses in tomato fruit. Physiol Plant 2013; 148: 445–455.

46. Romero LC, Aroca MÁ, Laureano-Marín AM, Moreno I, García I, Gotor C. Cysteine and cysteine-related signaling pathways in arabidopsis thaliana. Mol Plant 2014; 7: 264–276.

47. Gotor C, Laureano-Marín AM, Moreno I, Aroca Á, García I, Romero LC. Signaling in the plant cytosol: Cysteine or sulfide? Amino Acids 2014; 47: 2155–2164.

48. Turowski VR, Busi MV, Gomez-Casati DF. Structural and functional studies of the mitochondrial cysteine desulfurase from arabidopsis thaliana. Mol Plant 2012; 5: 1001–1010. [PubMed]

49. Kumar S, Kaur A, Chattopadhyay B, Bachhawat AK. Defining the cytosolic pathway of glutathione degradation in Arabidopsis thaliana: Role of the ChaC/GCG family of γ-glutamyl cyclotransferases as glutathione-degrading enzymes and AtLAP1 as the Cys-Gly peptidase. Biochem J 2015; 468: 73–85.

50. Biron DG, Brun C, Lefevre T, Lebarbenchon C, Loxdale HD, Chevenet F et al. The pitfalls of proteomics experiments without the correct use of bioinformatics tools. Proteomics 2006; 6: 5577–5596.

Index